Springer Theses

Recognizing Outstanding Ph.D. Research

Aims and Scope

The series "Springer Theses" brings together a selection of the very best Ph.D. theses from around the world and across the physical sciences. Nominated and endorsed by two recognized specialists, each published volume has been selected for its scientific excellence and the high impact of its contents for the pertinent field of research. For greater accessibility to non-specialists, the published versions include an extended introduction, as well as a foreword by the student's supervisor explaining the special relevance of the work for the field. As a whole, the series will provide a valuable resource both for newcomers to the research fields described, and for other scientists seeking detailed background information on special questions. Finally, it provides an accredited documentation of the valuable contributions made by today's younger generation of scientists.

Theses may be nominated for publication in this series by heads of department at internationally leading universities or institutes and should fulfill all of the following criteria

- They must be written in good English.
- The topic should fall within the confines of Chemistry, Physics, Earth Sciences, Engineering and related interdisciplinary fields such as Materials, Nanoscience, Chemical Engineering, Complex Systems and Biophysics.
- The work reported in the thesis must represent a significant scientific advance.
- If the thesis includes previously published material, permission to reproduce this must be gained from the respective copyright holder (a maximum 30% of the thesis should be a verbatim reproduction from the author's previous publications).
- They must have been examined and passed during the 12 months prior to nomination.
- Each thesis should include a foreword by the supervisor outlining the significance of its content.
- The theses should have a clearly defined structure including an introduction accessible to new PhD students and scientists not expert in the relevant field.

Indexed by zbMATH.

Shun Ito

Electronic Structures and Bonding Interaction of Structurally Defined Gold/Silver Superatoms

Probed by Anion Photoelectron Spectroscopy

Doctoral thesis accepted by
The University of Tokyo, Tokyo, Japan

Springer

Author
Dr. Shun Ito
Department of Chemistry
Graduate School of Science
The University of Tokyo
Tokyo, Japan

Supervisor
Prof. Tatsuya Tsukuda
Department of Chemistry
Graduate School of Science
The University of Tokyo
Tokyo, Japan

ISSN 2190-5053 ISSN 2190-5061 (electronic)
Springer Theses
ISBN 978-981-96-2891-9 ISBN 978-981-96-2892-6 (eBook)
https://doi.org/10.1007/978-981-96-2892-6

This Springer imprint is published by the registered company Springer Nature Singapore Pte Ltd.
The registered company address is: 152 Beach Road, #21-01/04 Gateway East, Singapore 189721, Singapore

Supervisor's Foreword

Gold and silver nanoclusters protected by organic ligands exhibit novel properties that cannot be predicted from the corresponding bulk due to unique atomic packing and quantized electronic structures. Their properties are influenced by various structural factors such as the number, element and arrangement of the constituent atoms of the Au/Ag core and the interaction between the core and the ligands. Recent advances in atomically precise synthesis, X-ray crystallography, and theoretical calculations have made it possible to more clearly correlate these structural factors with various properties and stabilities.

The key to understanding the origin of the novel properties and their design is to clarify how the electronic structures of ligand-protected Au/Ag nanoclusters are affected by the structural factors. Dr. Ito addressed this issue in his doctoral thesis by making full use of gas-phase anion photoelectron spectroscopy (PES) with density functional theory (DFT) calculations. The unique feature of anion PES is that the energy levels of the occupied orbitals can be probed without interference from the environment such as solvents and counterions. Therefore, the PE spectra can be directly assigned based on DFT calculations on optimized structures obtained from the reported crystal structures. He first developed a bright anion PE spectrometer and determined the electron affinity of the representative thiolate (RS)-protected Au cluster, $Au_{25}(SR)_{18}$, discovered in 2005. He then investigated the effects of heterometal doping and ligand substitution on the electronic structures. He rationalized the results in the framework of the "superatom concept", in which the valence electrons confined in a positive jellium potential provided by the Au/Ag core are accommodated in superatomic orbitals, such as 1S, 1P, 1D, etc. Finally, he evaluated the electronic structures of assemblies of superatoms with other superatoms or adatoms and explained them based on the bonding of superatoms.

Dr. Ito received the Young Best Presentation Award of the Society of Nano Science and Technology in 2021 and the CSJ Student Presentation Award in 2022 for his outstanding presentations. He also received the School of Science Encouragement Award from the University of Tokyo in 2019, 2021, and 2024 for his great academic and research achievements in bachelor's, master's, and doctoral studies, respectively.

I hope that this book will provide fundamental insights into the electronic structure of ligand-protected superatoms and pave the way for the rational development of superatom-based functional nanomaterials.

佃 達哉

Tokyo, Japan
December 2024

Prof. Tatsuya Tsukuda

List of published articles

Parts of this thesis have been published in the following journal articles:

1. Collision-Induced Reductive Elimination of 1,3-Diynes from $[MAu_{24}(C{\equiv}CR)_{18}]^{2-}$ (M = Pd, Pt) Yielding Clusters of Superatoms **S. Ito**, K. Koyasu, S. Takano, T. Tsukuda J. Phys. Chem. C **124**, 19119 (2020).
2. Critical Role of CF_3 Groups in the Electronic Stabilization of $[PdAu_{24}(C{\equiv}C_6H_3(CF_3)_2)_{18}]^{2-}$ as Revealed by Gas-Phase Anion Photoelectron Spectroscopy **S. Ito**, K. Koyasu, S. Takano, T. Tsukuda J. Phys. Chem. Lett. **12**, 10417 (2021).
3. Electron Affinities of Ligated Icosahedral M_{13} Superatoms Revisited by Gas-Phase Anion Photoelectron Spectroscopy **S. Ito**, Y. Tasaka, K. Nakamura, Y. Fujiwara, K. Hirata, K. Koyasu, T. Tsukuda J. Phys. Chem. Lett. **13**, 5049 (2022).
4. Correction to "Photoinduced Thermionic Emission from $[M_{25}(SR)_{18}]^-$ (M = Au, Ag) Revealed by Anion Photoelectron Spectroscopy" K. Hirata, **S. Ito**, K. Kim, K. Nakamura, H. Kitazawa, S. Hayashi, K. Koyasu, T. Tsukuda J. Phys. Chem. C **126**, 8965 (2022).
5. Supervalence Bonding in Bi-icosahedral Cores of $[M_1Au_{37}(SC_2H_4Ph)_{24}]^-$ (M = Pd and Pt): Fusion-Mediated Synthesis and Anion Photoelectron Spectroscopy E. Ito,* **S. Ito**,* S. Takano, T. Nakamura, T. Tsukuda JACS Au **2**, 2627 (2022). *These authors contributed equally to this work.
6. Resistance of a $PdAu_{12}$(8e) Core to Growth in Collision-Induced Sequential Reductive Elimination of $(C{\equiv}CR)_2$ from $[PdAu_{24}(C{\equiv}CR)_{18}]^{2-}$ **S. Ito**, K. Hirano, K. Koyasu, X.-K. Wan, Q.-M. Wang, T. Tsukuda J. Phys. Chem. Lett. **15**, 11060 (2024).

Acknowledgments The whole work of this doctoral thesis was conducted in the Department of Chemistry, the Graduate School of Science, the University of Tokyo under the supervision of Prof. Dr. Tatsuya Tsukuda.

First of all, I am profoundly grateful to my supervisor, Prof. Dr. Tatsuya Tsukuda. I deeply appreciate his great support for six years since I joined the group. He provided me a great environment for research and taught me not only the practical things such as the experimental technique, the writing of academic English, and the presentation skills but also the fundamental lessons such as the way of thinking and the behavior as a researcher and a leader. Without his patient support, I could give up my research on the way.

I am greatly obliged to Assoc. Prof. Dr. Kiichirou Koyasu for his instruction and discussion about the gas-phase experiment. I learned many techniques of gas-phase studies ranging from the maintenance and the design of the apparatus to the data analysis and theoretical calculations. His guidance was indispensable to conduct my study.

I would like to express my sincere gratitude to Asst. Prof. Dr. Shinjiro Takano for his assistance. He taught me the experimental techniques of synthesis of ligand-protected metal clusters in the solution especially when I was an undergraduate student. I appreciate his beneficial and inspiring advice in discussions. I would like to offer my special thanks to Asst. Prof. Dr. Shinya Masuda for discussions despite different research interests. Discussions with him were helpful to refresh my mind.

I am grateful to the Forefront Physics and Mathematics Program to Drive Transformation (FoPM), a World-leading Innovative Graduate Study (WINGS) Program at the University of Tokyo, for the opportunity to have interdisciplinary discussions and for the financial support for living expenses and the expenses for the international joint research in Germany. I really thank Prof. Dr. Fumio Komori and Prof. Dr. Kazumasa Takeuchi for fruitful discussions as secondary supervisors. I would like to deeply thank Prof. Dr. Manfred Kappes (Karlsruhe Institute of Technology, Germany) for accepting my research stay in his laboratory for three months.

I would like to appreciate the Research Center for Computational Science (National Institute of National Science) for providing the calculation resource of density functional theory calculation and appreciate the research fellowships and grants from Japan Society for the Promotion of Science and Japan Science and Technology Agency.

Finally, I would like to be deeply grateful to my family and the laboratory members for their kind support.

Shun Ito

Competing Interests The author has no competing interests to declare that are relevant to the content of this manuscript.

Contents

Chapter 1
Introduction

Abstract The first chapter introduces the overall background of my doctoral thesis. Metal clusters and the concept of superatoms are introduced based on the gas-phase studies. Ligand protected clusters are then introduced and conceptualized as structurally defined superatoms. The theoretical principle of anion photoelectron spectroscopy intensively used in the thesis is also introduced. Finally, I remark the aim and the structure of the thesis.

Keywords Metal clusters · Ligand-protected gold clusters · Superatoms · Anion photoelectron spectroscopy

1.1 Metal Clusters

Metal clusters are substances composed of a few to tens of metal atoms. Metal clusters show novel properties completely different from those of bulk metals due to the unique atomic packing and the discrete energy levels (Fig. 1.1). Metal clusters can take unique geometries different from typical closed-packed structures to suppress the surface energy. Continuous band structures of bulk metals and nanoparticles are quantized in metal clusters due to the decrease of electronic density of states, so-called a quantum size effect. These geometrical and electronic features make metal clusters fascinating materials: the properties of metal clusters are not scalable from those of bulks or atoms at all and the difference of single atom can make a drastic change in their properties.

Due to the intrinsic instability against aggregation and the need of size-selected characterization, naked metal clusters have been prepared and thoroughly characterized in a vacuum chamber by using mass spectrometry and highly sensitive gas-phase spectroscopy. An intriguing example of the size-specific property is mass spectrometry of sodium clusters by W. D. Knight in 1984 [1]. Sodium clusters were generated by vaporizing bulk sodium at high temperature of about 900 °C and the mass spectrum of the product showed that sodium clusters with the size of 8, 20, 40, 58, and 92 were highly abundant compared to other species. Another famous example of the

S. Ito, *Electronic Structures and Bonding Interaction of Structurally Defined Gold/Silver Superatoms*, Springer Theses, https://doi.org/10.1007/978-981-96-2892-6_1

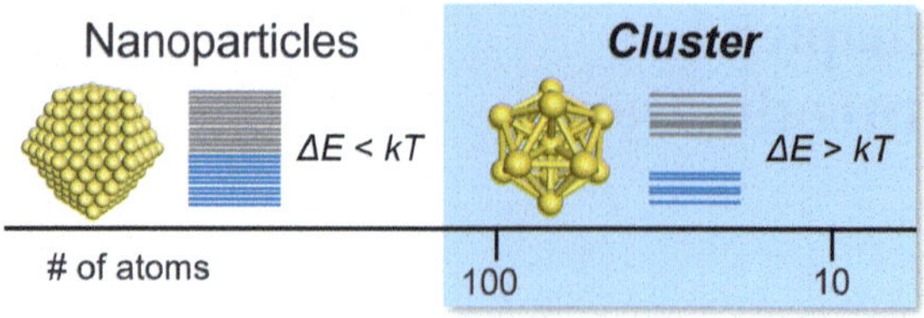

Fig. 1.1 Conceptual illustration of nanoparticles and clusters

size-specific stability is the mass spectrometric study of the reaction of aluminum clusters with oxygen gas [2]. Without oxygen gas various sizes of aluminum clusters were observed. However, after the reaction with oxygen gas Al_{13}^- and Al_{23}^- became prominent while other aluminum clusters went away. Size-specific properties were found also in the electronic structure as demonstrated in the photoelectron spectrum of Au_{20}^- [3]. The peak X was assigned to the transition from Au_{20}^- to the ground state of neutral Au_{20}^0 while the peak A is to the first excited state of neutral Au_{20}^0. Therefore, the gap between the peak A and X represented the energy gap between highest occupied molecular orbital (HOMO) and lowest unoccupied molecular orbital (LUMO) of neutral Au_{20}^0. The HOMO–LUMO gap of Au_{20}^0 thus calculated was 1.77 eV, which is greater than that of fullerene C_{60} (1.57 eV), the most famous carbon cluster, and implying the potential of practical application under ambient condition.

The size-specific stabilities and properties of metal clusters can be explained by a spherical jellium model [1, 4, 5]. In the spherical jellium model, the valence electrons of metal clusters behave like a free electron gas confined in a positive jellium (jelly-like) potential with a finite depth from the vacuum level (Fig. 1.2a). The shape of the jellium potential is often described by a Woods-Saxon potential [6], which has successfully explained the magic numbers of nucleons in the field of nuclear physics. The Woods-Saxon potential is described in terms of the distance r from the center of metal clusters:

$$E(r) = -\frac{U}{1+\exp\left(\frac{r-r_0}{\varepsilon}\right)} \tag{1.1}$$

where U, r_0, and ε denote the depth of the potential, the radius of metal clusters, and the sharpness of the edge of the potential, respectively. The solution of Schrödinger equation with the Woods-Saxon potential shows that an electron in the Woods-Saxon potential takes discrete energy levels (Fig. 1.2b) and the probability densities of the electron at each energy level resemble that of atomic orbitals (Fig. 1.2c). Therefore, these energy levels are called 1S, 1P, 1D, 2S by using a radial quantum number n' indicating the number of nodes of wave functions along the radial direction and the angular momentum quantum number l where S, P, D, F is corresponding to $l = 0, 1, 2, 3$. It is noteworthy that the meaning of the number in 1S, 1P, 1D, 2S is different from that of atomic orbitals 1s, 2s, 2p, 3s while the meaning of the character is the same: the number in atomic orbitals means the principal quantum number n obtained by the sum of the radial quantum number and the angular momentum quantum number:

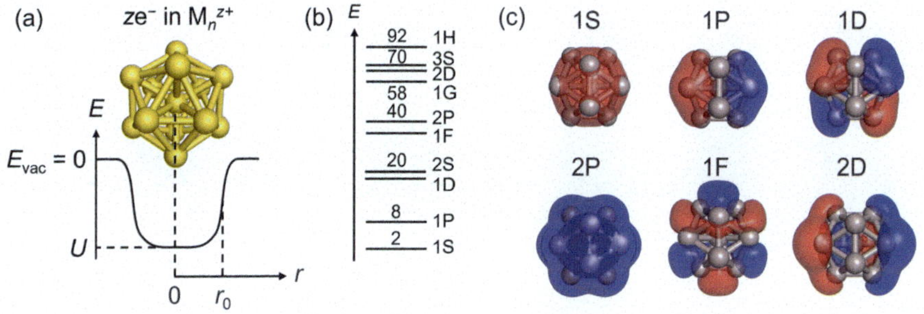

Fig. 1.2 **a** schematic illustration of the jellium potential of metal clusters. **b** Energy diagram for a spherical Woods-Saxon potential. Numbers above or below the level indicate the number of electrons required to fill up the level. **c** The shape of superatomic orbitals obtained by density functional theory calculation of Al_{13}—Cluster. The calculation level was B3LYP functional and 6-31G(d) basis set. Contour values were set to 0.015 in all orbitals

$$n = n' + l. \tag{1.2}$$

Therefore, 1S, 1P, 1D, 2S orbitals are corresponding to 1s, 2p, 3d, 2s orbitals. The order of energy levels is different due to the different shape of the attractive potential: the atomic nucleus behaves like a point charge to produce a Coulombic potential but the positive charge of metal clusters smeared over the entire cluster to make a jellium potential.

The spherical jellium model plays an important role in the stabilities and properties. In case of sodium clusters, each sodium atom provides one free electron from 3s orbital and the magic numbers of sodium clusters 8, 20 are corresponding to the closed-shell electron configuration of $(1S)^2(1P)^6$ and $(1S)^2(1P)^6(1D)^{10}(2S)^2$ (Fig. 1.2b). To describe the number of valence electrons in the cluster, the following notation is used in my thesis: Na_8(8e) and Na_{20}(20e). The magic stability of Al_{13}^- and Al_{23}^- against the oxygen etching was ascribed to the closed-shell electron configuration with Al_{13}^-(40e) up to 2P and Al_{23}^-(70e) up to 3S (Fig. 1.2b) because each aluminum atom provides three free electrons from $(3s)^2(3p)^1$. These examples showed that metal clusters gain high stability and behave like inert rare gas atom when they take the closed-shell electron configuration. The large HOMO–LUMO gap of neutral Au_{20} was ascribed to the closed-shell electron configuration with Au_{20}(20e) up to 2S orbitals and the large energy gap between superatomic 2S and 1F orbitals (Fig. 1.2b).

The similarity of the electronic shell, the orbital shape, and the chemical properties between metal clusters and atoms led to a **superatom** concept [7–12] and the energy levels 1S, 1P, 1D are called superatomic orbitals. Al_{13}^-(40e) and Au_{20}(20e) with closed-shell electron configuration corresponded to inert rare-gas atoms He, Ne, Ar while Al_{13}^0(39e) was categorized to a superhalogen based on a large electron affinity of 3.57 eV exceeding that of halogen atoms and the formation of an ionically bound cluster $Al_{13}I^-$ [13]. Superatoms have been expected to compose a novel three-dimensional periodic tables with the size and composition as the third dimension and

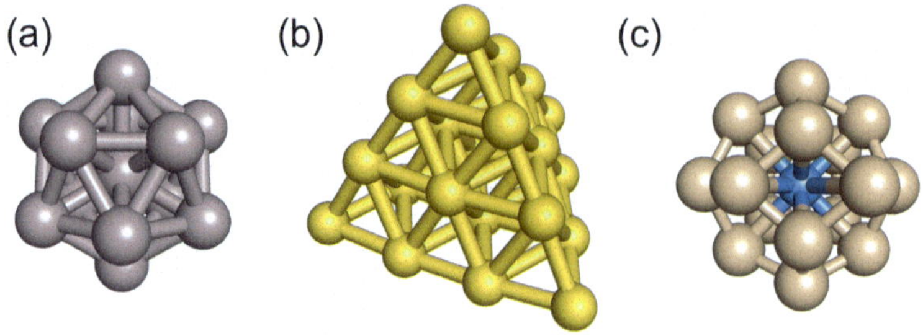

Fig. 1.3 Geometrical structures of **a** Al_{13}^-, **b** Au_{20}^0, and **c** Ta@Si_{16}

expected to be novel building blocks of materials. Based on this idea, several stable superatoms such as Al_{13}^-(40e), Au_{20}^0(20e), M@Si_{16}(68e) (M = Sc^-, Ti^0, V^+, Y^-, Zr^0, Nb^+, Lu^-, Tb^0, Ho^+, Hf^0, Ta^+) [14] were found to be a promising candidate (Fig. 1.3), and the generation of superatom-based materials has been intensively studied [15–17].

1.2 Ligand-Protected Superatoms

Apart from gas-phase studies of metal clusters, gold clusters protected by organic ligands have been synthesized in solution since 1965 [18]. At the initial era, components were limited to only several species such as gold, halide, and phosphine. In recent two decades the field of ligand-protected gold/silver clusters has grown rapidly due to mainly three reasons: (1) the improvement of the stability by the discovery of adequate ligands such as thiolate [19], alkynyl [20], diphosphine [21], and N-heterocyclic carbenes [22]; (2) the development of easy and reproducible wet chemical synthesis at atomic precision; (3) the unambiguous determination of geometrical structures by single crystal X-ray crystallography (SCXRD).

The representative and most intensively studied gold cluster is $[Au_{25}(SR)_{18}]^-$ (Fig. 1.4) because of its early discovery, high stability, easy preparation, diverse functionalization, and potential application [23]. Here I briefly overview the history of $[Au_{25}(SR)_{18}]^-$. The cluster was firstly synthesized by R. W. Murray in 2004 by the reduction of Au(I)-PET (PET = 2-PhC_2H_4S) complex with a strong reducing agent $NaBH_4$ in a two-phase system [24]. The composition was firstly assigned to $Au_{38}(PET)_{24}$ based on the thermogravimetric analysis (TGA) and elemental analysis, but later corrected as $[Au_{25}(PET)_{18}]^-$ based on the electrospray ionization mass spectrometry (ESI–MS) [25]. The same composition was also observed in ESI–MS of the series of glutathione (SG)-protected clusters $Au_x(SG)_y$ separated by polyacrylamide gel electrophoresis (PAGE) [26]. Several geometries of $[Au_{25}(PET)_{18}]^-$ were proposed based on density functional theory (DFT) calculation [27, 28], but the structure determination by SCXRD of the salt with tetraoctylammonium cation ($(C_8H_{17})_4N^+$) was reported in 2008 independently from Murray [29] and Jin [30].

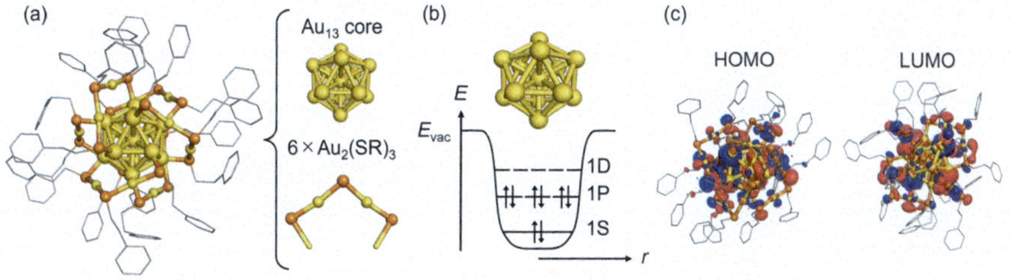

Fig. 1.4 **a** Crystal structure, **b** schematic superatomic electronic structure, and **c** typical Kohn–Sham orbitals of $[Au_{25}(PET)_{18}]^-$ (PET = 2-PhC_2H_4S)

Two reported structures were identical as shown in Fig. 1.4a and the structure can be divided into two parts: an icosahedral Au_{13} core and six protecting oligomers $Au_2(PET)_3$. The unique properties of $[Au_{25}(SR)_{18}]^-$ such as the optical [31], chiral [32], magnetic [33], electrochemical [34], and catalytic properties [35] were thoroughly investigated as a benchmark of ligand-protected clusters. $[Au_{25}(SR)_{18}]^-$ have been functionalized by using various metals and ligands. The representative examples are a heterometal-doped cluster $[PtAu_{24}(SR)_{18}]^-$ [36], a silver-based cluster $[Ag_{25}(SR)_{18}]^-$ [37], an alkynyl-protected cluster $[Au_{25}(C{\equiv}CR)_{18}]^-$ [38].

Although ligand-protected clusters have been developed in a different viewpoint from naked metal clusters, it was reported that ligand-protected gold clusters can be also viewed as superatoms [39]. The formal number of valence electrons N of ligand-protected gold clusters $[Au_xL_y]^{z+}$ can be counted by the following formula

$$N = x - y - z \tag{1.3}$$

where x, y, and z denotes the number of gold atoms, the number of electron-withdrawing ligands such as thiolate (SR), alkynyl (C≡CR), and halide, and the net charge of the cluster, respectively, while neutral ligands such as phosphine and N-heterocyclic carbene does not contribute to the electron counting. It was demonstrated that ligand-protected gold clusters with spherical cores gain high stability when the number of valence electrons fulfills the closed-shell electron configuration of superatoms (Fig. 1.2b) [39]. For example, the number of valence electrons of the representative cluster $[Au_{25}(SR)_{18}]^-$ is calculated to be $8(= 25 - 18 - (-1))$ and corresponding to the closed-shell electron configuration with $(1S)^2(1P)^6$ and referred to as $[Au_{25}(SR)_{18}]^-$(8e) (Fig. 1.4b).

The superatom concept has explained the stability of the various type of ligand-protected clusters [40–42]. $[Au_9(PPh_3)_8]^{3+}$(6e) [43, 44] with a prolate Au_9 core (Fig. 1.5a) looks unstable in terms the electron counting. However, triply degenerated 1P orbitals were separated into two stable 1P orbitals and one unstable 1P orbital by a Jahn–Teller effect of the prolate Au_9 core and gained a high stability with $(1S)^2(1P)^4$ configuration. Ligand-protected clusters with different metals can be also discussed in a superatom concept by assuming group 11 elements (Ag, Cu) provide one valence

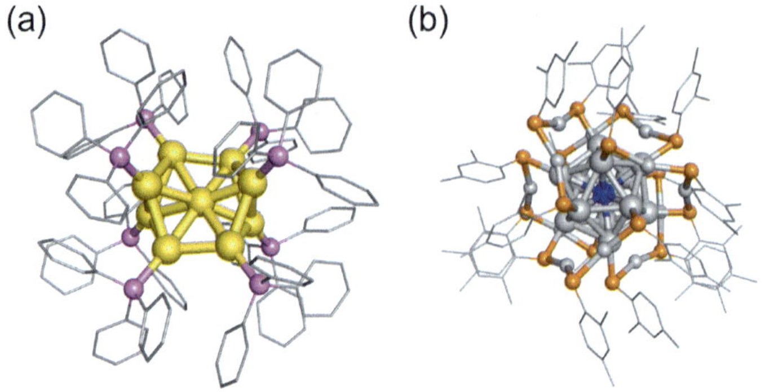

Fig. 1.5 Crystal structures of **a** $[Au_9(PPh_3)_8]^{3+}$(6e), and **b** $[PdAg_{24}(DMBT)_{18}]^{2-}$(8e) (DMBT = 2,4-$(CH_3)_2C_6H_3$)

electron while group 10 elements (Pd, Pt, Ni) does not provide any valence electrons. In this scheme, both silver-based clusters $[Ag_{25}(DMBT)_{18}]^-$ (DMBT = 2,4-$(CH_3)_2C_6H_3$)[37] and $[MAg_{24}(DMBT)_{18}]^{2-}$ (M = Pd, Pt) [45] (Fig. 1.5b) are also eight electron superatoms with $(1S)^2(1P)^6$ configuration. In my doctoral thesis, ligand-protected gold/silver clusters with superatomic characters and unambiguous geometries identified by SCXRD are defined as **structurally defined superatoms**. The advantageous points of structurally defined superatoms compared to naked metal superatoms are (1) the high stability owing to ligands (2) precise structural modification by wet chemical methods (3) the structure–property relationship based on the well-defined structure. Thus, we can develop discussions about its properties based on the well-established geometries of structurally defined superatoms.

It has been reported that multiple superatoms can be directly connected via Au–Au bonding. For example, $Au_{38}(PET)_{24}$(14e) [46] (Fig. 1.6a) has an elongated Au_{23} core composed of two Au_{13} icosahedrons by sharing the triangle Au_3 facet. Theoretical chemists proposed a super valence bond (SVB) model to explain the stability of the elongated Au_{23} core [47, 48]. In the SVB model, Kohn–Sham orbitals of $Au_{23}{}^{9+}$ can be viewed as the σ and π bonds between 1S and 1P orbitals at each Au_{13} core (Fig. 1.6b). The electron configuration of Au_{23} core with 14 electrons was expressed as $(1\Sigma)^2(1\Sigma^*)^2(1\Pi)^4(2\Sigma)^2(1\Pi^*)^4$ where capital letters were used to distinguish the supervalence bonding of superatoms from the valence bonding of conventional molecules. Thus, the electron configuration of the Au_{23} core was an exact analogue of the F_2 molecule and $Au_{38}(PET)_{24}$ can be viewed as superatomic molecules.

The superatom concept plays an important role not only in the stability but also in the property, because both HOMO and LUMO of structurally defined superatoms are often superatomic orbitals. The shape of HOMO and LUMO of $[Au_{25}(SR)_{18}]^-$ is shown in Fig. 1.4c. Based on the symmetry analysis of molecular orbitals, HOMO and LUMO are assigned to superatomic 1P and 1D orbitals while some of the electrons were delocalized over the protecting Au–S oligomers [39]. Therefore, these superatomic orbitals are deeply correlated with the chemical properties relating to frontier

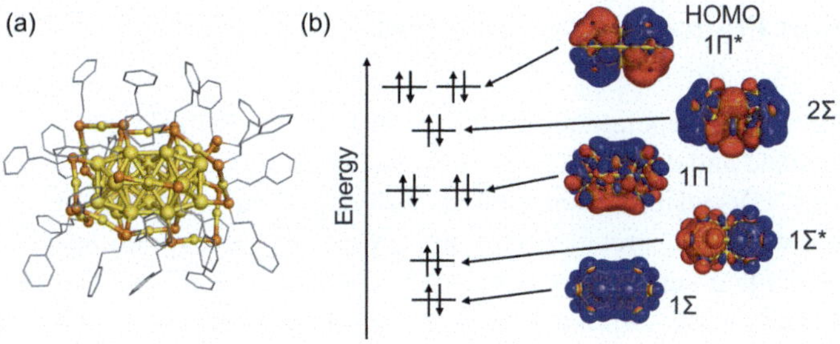

Fig. 1.6 **a** Crystal structure of $Au_{38}(PET)_{24}$(14e) (PET = 2-PhC_2H_4S). **b** Kohn–Sham orbitals of Au_{23}^{9+} having 14 valence electrons

orbitals such as optical properties and electron transfer processes (Fig. 1.7ab). The UV–visible absorption spectrum of $[Au_{25}(SR)_{18}]^-$ of solution (Fig. 1.7c) exhibits the distinct peak around 1.8 eV, which was assigned to the HOMO–LUMO transition where an electron in 1P orbitals is excited to 1D orbitals [30]. The superatomic electron configuration of the lowest excited state is $(1S)^2(1P)^5(1D)^1$ and the photoluminescence of $[Au_{25}(SR)_{18}]^-$ is also assigned to the relaxation of an electron from 1D orbitals to 1P orbitals. Electrochemical measurement showed that the first oxidation potential and the first reduction potential of $[Au_{25}(SC_6H_{13})_{18}]^-$ were determined to be 0.11 eV and −1.55 eV versus the ferrocenium/ferrocene couple [34]. These potentials involve the electron transfer from 1P orbitals and that to 1D orbitals, respectively. The intimate relationship between the property and the superatomic orbitals motivated the fundamental study of the electronic structure of structurally defined superatoms in my thesis.

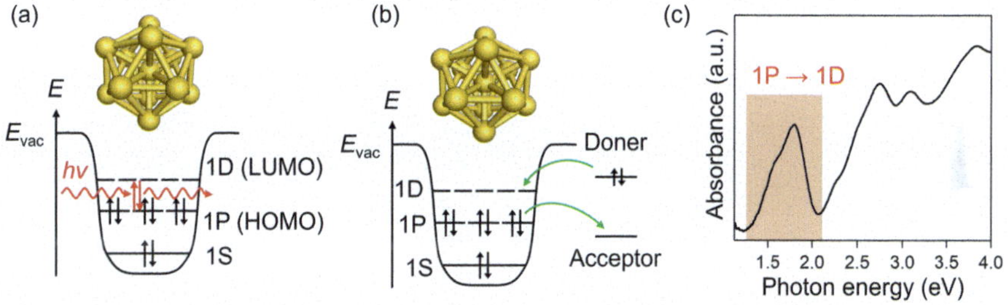

Fig. 1.7 **a** Schematic illustration of fundamental roles of superatomic orbitals in **a** optical properties and **b** electron transfer processes. **c** UV–visible absorption spectrum of $[Au_{25}(PET)_{18}]^-$

1.3 Anion Photoelectron Spectroscopy

The evaluation of the superatomic orbitals of structurally defined superatoms is essential to understand and design its properties. Such characterizations have been mainly conducted by UV–visible absorption spectroscopy and electrochemical measurement as shown in the previous section. UV–visible absorption spectroscopy of structurally defined superatoms is not only convenient and easy but also used as fingerprints because absorption spectra are sensitive to its composition and structures [49]. Electrochemical measurement has revealed the redox properties of structurally defined superatoms in solution and utilized in the design of the electrocatalysts and electrochemical sensors [34]. However, the intrinsic electronic structure is still elusive because both methods are conducted in the solution and perturbated by the environments such as solvents and counter ions [50].

Another method to evaluate the electronic structure of substances is **photoelectron spectroscopy (PES)** [51]. In PES, a light with a known energy $h\nu$ is irradiated on substances and photoelectrons e^- are detached by the photoelectric effect (Fig. 1.8a). The combination of photons (UV, VUV, X-ray) and substances (gas, solid, droplet) is diverse, but here I utilized gas-phase anion PES using UV light to evaluate the electronic structure of structurally defined superatoms. The advantage of anion PES is that (1) we can elucidate the electronic structure around HOMO at a high resolution, (2) spectroscopy is conducted under an ideal condition without interferences from environments, and (3) the obtained spectrum is directly comparable to the result of electronic density of states obtained by DFT calculations. Anion PES was firstly developed by Smalley [52] to investigate the electronic structure of metal clusters in the gas phase and successfully illustrated the size-specific properties of metal clusters.

The principle of PES of an anion M^- is described by the following chemical equation

$$M^- + h\nu \rightarrow M^0 + e^- \tag{1.4}$$

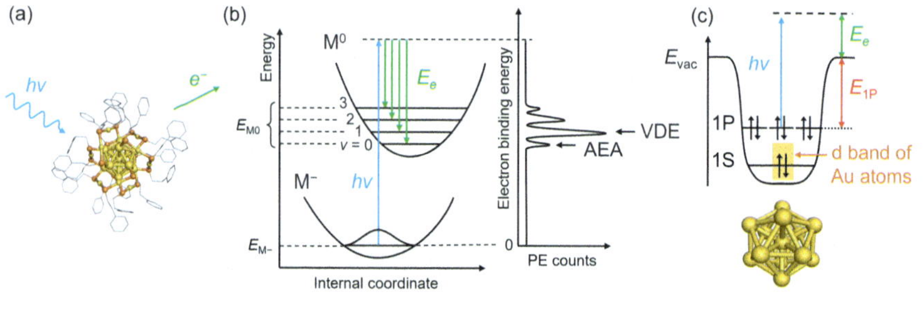

Fig. 1.8 **a** Illustration of anion photoelectron spectroscopy, **b** scheme of the principle of anion photoelectron spectroscopy, and **c** schematic illustration of anion photoelectron spectroscopy of structurally defined superatoms

and the total energy of the system is conserved during the process

$$E_{M^-} + h\nu = E_{M^0} + E_e \quad (1.5)$$

where E_{M^-}, E_{M^0}, and E_e denote the energy of the anion, the corresponding neutral, and the detached photoelectron, respectively. Equation (1.5) can be transformed to the following equation

$$E_{M^0} - E_{M^-} = h\nu - E_e \quad (1.6)$$

and Eq. (1.6) shows that the energy difference between anions and neutrals, defined as an electron binding energy, can be determined by measuring the kinetic energy of photoelectrons.

Most of the anions are in the vibrationally ground state, but the final state of neutrals can be vibrationally excited state (Fig. 1.8b). The intensity of photoelectron signals regarding each excited state is explained by a Franck–Condon principle [53]: the transition probability is proportional to the square of the overlap integral between the vibrational functions of anions and neutrals assuming the Born–Oppenheimer approximation (wave functions of nuclei and electrons can be treated separately). Thus, the most intense signal among the vibrational progression is the vertical transition to the vibrational state having the largest overlap integral with anions at the equilibrium structure of anions. The vibronic structure in the PE spectrum tells us how the geometrical structure of the neutral is changed from that of the anion. The energy required for the vertical transition is called as **vertical detachment energy (VDE)**. On the other hand, the transition with the lowest energy is called as the adiabatic transition and the required energy is called as **adiabatic electron affinity (AEA)** of corresponding neutrals. The electron binding energy obtained by anion PES can be correlated with the energy of molecular orbitals under the Koopmans' theorem, where the nth ionization energy of molecules is equal to the negative value of the energy of the nth occupied molecular orbitals from HOMO by assuming the molecular orbitals of neutrals are the same with those of anions.

The schematic illustration of the anion PES of structurally defined superatoms is shown in Fig. 1.8c. The electrons in superatomic 1P orbitals are excited by photons and detached as photoelectrons. The electron binding energy of superatomic 1P orbitals can be obtained by subtracting the kinetic energy of photoelectrons from the energy of photons. Anion PES is especially suitable to evaluate 1P orbitals located at HOMO, while the energy level of 1S orbital is difficult to evaluate because it is often buried in the 5d band of Au atoms. It should be noted that the vibrational progression does not resolve clearly in case of complex molecules including clusters due to the dense vibrational states and the complicated potential energy surfaces. Therefore, VDE and AEA is defined as the peak position and the onset of the peak, respectively. The experimental methodology of anion PES is summarized in Chap. 2.

In case of anion PES of multiply charged anions, electrostatic repulsion between the detached photoelectrons and remaining anions must be considered to interpret the PE spectrum correctly [54]. Let us consider an electron e^- approaching to a monoanion M^- with a positive electron affinity (= a positive second electron affinity of corresponding neutral). The electron will be captured by the monoanion when they are close enough. However, there is a long-range repulsive Coulomb force between the electron and the monoanion, which acts as a barrier for approaching electrons and is called as a **repulsive Coulomb barrier (RCB)**. An electron with zero or a small kinetic energy cannot reach the monoanion due to the RCB. The RCB also exists in case of the detachment of electrons from dianions M^{2-}. The potential energy surfaces shown in Fig. 1.9 illustrate the principle of PES of dianions with the third axis: the distance between the detached photoelectrons and remaining anions. Photoelectrons must overcome the RCB to be emitted, while the final kinetic energy of photoelectrons is determined by the energy difference between the monoanion and the dianions as in the case of PES of monoanions (Fig. 1.8b). In other words, photoelectrons with less kinetic energies than the height of RCB cannot detach from dianions. The effect of the RCB is observed as the rapid decrease of photoelectron counts at the high electron binding energy. Representative example of the RCB is the PE spectra of a quadruply charged anion, Copper phthalocyanine sulfonic acid $[CuPc(SO_3)_4]^{4-}$ [54]. The PE spectrum recorded at 193 nm (6.42 eV) exhibited a peak A at the electron binding energy of 1 eV and a peak B at 2 eV. In the PE spectrum recorded at 266 nm (4.66 eV), however, PE counts were scarce at the high electron binding energy beyond 1 eV and the peak B disappeared. Scarce PE signals at the high electron binding energy at 266 nm were ascribed to the cutoff of PEs with small kinetic energies by the RCB. The height of the RCB was estimated by subtracting the apparent position of the peak A at 266 nm from the photon energy. PES of multiply charged anions demonstrated the unique phenomena such as a negative electron affinity of highly negatively-charged anions and a tunneling electron detachment via photoexcited states through the RCB [55, 56].

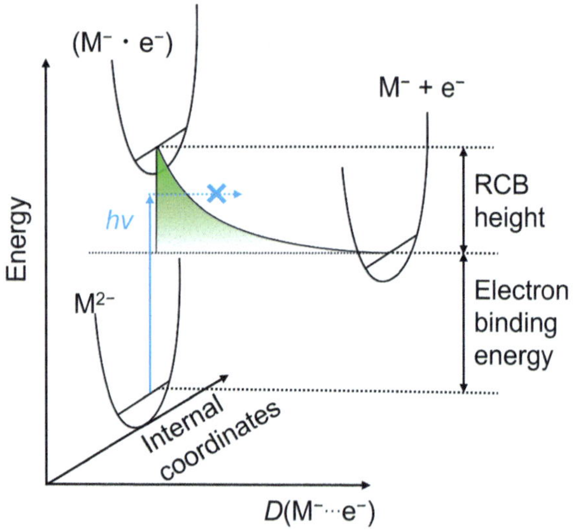

Fig. 1.9 Schematic illustration of anion photoelectron spectroscopy of multiply charged anions

1.4 Aim of My Study

As discussed in Sect. 1.2, ligand-protected gold/silver clusters can be categorized as structurally defined superatoms. Superatomic orbitals of structurally defined superatoms are closely related to its property and the intrinsic evaluation of superatomic orbitals can facilitate the understanding and the rational design of functional superatoms. In my thesis, I addressed a fundamental question that how the electronic structure of superatomic orbitals is affected by the structural modification and how multiple superatoms make chemical bonding with each other as superatomic molecules. To this end, I utilized anion PES introduced in Sect. 1.3 to evaluate the superatomic orbitals around HOMO.

My doctoral thesis is composed of one chapter describing the experimental setup of anion PES and four chapters describing the results and discussions. Chaps. 3 and 4 focus on the effect of the structural modification on the superatomic orbitals, while Chaps. 5 and 6 summarize the bonding interaction between superatoms. In Chap. 2, the principle and the setup of the experimental methodology of anion PES used in Chaps. 3–6 is described. In Chap. 3, the result of anion PES of the representative superatom $[Au_{25}(PET)_{18}]^-$(8e) (PET = 2-PhC_2H_4S) and its interpretation is presented and the effect of heterometal doping on 1P orbitals is discussed based on anion PES of $[MAg_{24}(DMBT)_{18}]^z$(8e) (DMBT = 2,4-$(CH_3)_2C_6H_3S$; $z = -1$ for M = Ag, Au; $z = -2$ for M = Pd, Pt). In Chap. 4, the effect of stepwise ligand substitution of the energy level of 1P orbitals is investigated by using $[PdAu_{24}(C{\equiv}CAr^F)_{18-x}(C{\equiv}CPh)_x]^{2-}$(8e) ($Ar^F$ = 3,5-$(CF_3)_2C_6H_3$; x = 0–6). In Chap. 5, the bonding interaction between two superatoms is investigated by anion PES of chemically synthesized homonuclear superatomic molecules $[MAu_{37}(PET)_{24}]^-$ (14e) (M = Pd, Pt). In Chap. 6, the novel heteronuclear superatomic molecules are discovered in $[PdAu_{24}(C{\equiv}CAr^F)_{18-2n}]^{2-}$((8 + 2$n$)e) ($n$ = 0–6) generated by the collision-induced reductive elimination of $[PdAu_{24}(C{\equiv}CAr^F)_{18}]^{2-}$(8e) in the gas phase. In Chap. 7, I described the overall summary of the thesis and provided future prospects.

References

1. Knight WD, Clemenger K, de Heer WA, Saunders WA, Chou MY, Cohen ML (1984) Phys Rev Lett 52:2141
2. Leuchtner RE, Harms AC, Castleman AW (1989) J Chem Phys 91:2753
3. Li J, Li X, Zhai H-J, Wang L-S (2003) Science 299:864
4. de Heer WA (1993) Rev Mod Phys 65:611
5. Brack M (1993) Rev Mod Phys 65:677
6. Woods RD, Saxon DS (1954) Phys Rev 95:577
7. Khanna SN, Jena P (1995) Phys Rev B 51:13705–13716
8. Luo Z, Castleman AW (2014) Acc Chem Res 47:2931–2940
9. Reber AC, Khanna SN (2017) Acc Chem Res 50:255–263
10. Doud EA, Voevodin A, Hochuli TJ, Champsaur AM, Nuckolls C, Roy X (2020) Nat Rev Mater 5:371

11. Yin B, Luo Z (2021) Coord Chem Rev 429:213643
12. Yu F, Li J, Liu Z, Wang R, Zhu Y, Huang W, Liu Z, Wang Z (2023) J Clust Sci 34:1691
13. Bergeron DE, Castleman AW Jr, Morisato T, Khanna SN (2004) Science 304:84
14. Koyasu K, Akutsu M, Mitsui M, Nakajima A (2005) J Am Chem Soc **127:**4998
15. Zhang Q-F, Chen X, Wang L-S (2018) Acc Chem Res 51:2159
16. Tsunoyama H, Shibuta M, Nakaya M, Eguchi T, Nakajima A (2018) Acc Chem Res 51:1735
17. Inoue T, Hatanaka M, Nakajima A (2023) J Am Chem Soc 145:23088
18. Malatesta L, Naldini L, Simonetta G, Cariati F (1965) Chem Commun 11:212
19. Kawawaki T, Ebina A, Hosokawa Y, Ozaki S, Suzuki D, Hossain S, Negishi Y (2021) Small 17:2005328
20. Zhang M-M, Dong X-Y, Wang Y-J, Zang S-Q, Mak TCW (2022) Coord Chem Rev 453:214315
21. Adnan RH, Madridejos JML, Alotabi AS, Metha GF, Andersson GG (2022) Adv Sci 9:2105692
22. Narouz MR, Osten KM, Unsworth PJ, Man RWY, Salorinne K, Takano S, Tomihara R, Kaappa S, Malola S, Dinh C-T et al (2019) Nat Chem 11:419
23. Kang X, Chong H, Zhu M (2018) Nanoscale 10:10758
24. Donkers RL, Lee D, Murray RW (2004) Langmuir 20:1945
25. Tracy JB, Crowe MC, Parker JF, Hampe O, Fields-Zinna CA, Dass A, Murray RW (2007) J Am Chem Soc 129:16209
26. Negishi Y, Nobusada K, Tsukuda T (2005) J Am Chem Soc 127:5261
27. Iwasa T, Nobusada K (2007) J Phys Chem C 111:45
28. Akola J, Walter M, Whetten RL, Häkkinen H, Grönbeck H (2008) J Am Chem Soc 130:3756
29. Heaven MW, Dass A, White PS, Holt KM, Murray RW (2008) J Am Chem Soc 130:3754
30. Zhu M, Aikens CM, Hollander FJ, Schatz GC, Jin R (2008) J Am Chem Soc 130:5883
31. Aikens CM (2018) Acc Chem Res 51:3065
32. Li Y, Higaki T, Du X, Jin R (2020) Adv Mater 32:1905488
33. Agrachev M, Ruzzi M, Venzo A, Maran F (2019) Acc Chem Res 52:44
34. Kwak K, Lee D (2019) Acc Chem Res 52:12
35. Du X, Jin R (2019) ACS Nano 13:7383
36. Qian H, Jiang DE, Li G, Gayathri C, Das A, Gil RR, Jin R (2012) J Am Chem Soc 134:16159
37. Joshi CP, Bootharaju MS, Alhilaly MJ, Bakr OM (2015) J Am Chem Soc 137:11578
38. Li J-J, Guan ZJ, Lei Z, Hu F, Wang Q-M (2019) Angew Chem Int Ed **58:**1083
39. Walter M, Akola J, Lopez-Acevedo O, Jadzinsky PD, Calero G, Ackerson CJ, Whetten RL, Gronbeck H, Hakkinen H (2008) Proc Natl Acad Sci 105:9157
40. Hirai H, Ito S, Takano S, Koyasu K, Tsukuda T (2020) Chem Sci 11:12233
41. Omoda T, Takano S, Tsukuda T (2021) Small 17:2001439
42. Takano S, Tsukuda T (2021) J Am Chem Soc 143:1683
43. Wen F, Englert U, Gutrath B, Simon U (2008) Eur J Inorg Chem 2008:106
44. Yamazoe S, Matsuo S, Muramatsu S, Takano S, Nitta K, Tsukuda T (2017) Inorg Chem 56:8319
45. Yan J, Su H, Yang H, Malola S, Lin S, Häkkinen H, Zheng N (2015) J Am Chem Soc 137:11880
46. Qian H, Eckenhoff WT, Zhu Y, Pintauer T, Jin R (2010) J Am Chem Soc 132:8280
47. Cheng L, Yang J (2013) J Chem Phys 138:141101
48. Cheng L, Ren C, Zhang X, Yang J (2013) Nanoscale 5:1475
49. Jin R (2015) Nanoscale 7:1549
50. Kang X, Chen S, Jin S, Song Y, Xu Y, Yu H, Sheng H, Zhu M (2016) ChemElectroChem 3:1261
51. Eland JHD (1984) Photoelectron spectroscopy. Butterworth, London
52. Cheshnovsky O, Yang SH, Pettiette CL, Craycraft MJ, Smalley RE (1987) Magn Rev Sci Instrum 58:2131
53. Franck J, Dymond EG (1926) Trans Faraday Soc 21:536
54. Wang L-S, Wang X-B (2000) J Phys Chem A 104:1978
55. Wang X-B, Ding CF, Wang L-S (1998) Phys Rev Lett 81:3351
56. Wang X-B, Wang L-S (1999) Nature 400:245

Chapter 2
Experimental Setup of Anion Photoelectron Spectroscopy

Abstract The second chapter describes the experimental setup of gas-phase anion photoelectron spectroscopy. The composition of the apparatus is overviewed in the first section. Details of each component (an electrospray ionization source, a quadrupole linear ion trap, a time-of-flight mass spectrometer, and a magnetic-bottle photoelectron spectrometer) are described in the following sections. Especially, the quadrupole linear ion trap I designed and implemented is described in detail.

Keywords Anion photoelectron spectroscopy · Electrospray ionization · Mass spectrometry · Vacuum experiment · Ion trap

2.1 Overview of Apparatus

A gas-phase anion photoelectron spectrometer consists of four main components: an electrospray ionization (ESI) source, a quadrupole linear ion trap (QLIT), a time-of-flight mass spectrometer (TOF–MS), and a magnetic-bottle photoelectron spectrometer (MB-PES). 2D and 3D drawings of the apparatus are shown in Figs. 2.1, 2.2, and 2.3. An ESI source and a TOF–MS was constructed by Tsukuda and Negishi about two decades ago [1, 2]. A MB-PES was installed by Hirata in 2017 [3]. A QLIT was newly designed and installed by me in 2021.

The typical procedure of the experiment is as follows. Firstly, structurally defined superatoms were synthesized at atomic precision by a wet chemical synthesis. Then, the solution of structurally defined superatoms was ionized by the ESI source and generated anions were introduced into the vacuum chamber through a desolvation capillary. Continuously generated ions were accumulated in the QLIT to produce intensified ion pulses. Ions were released from the QLIT and mass-selected by the TOF–MS. Spatially and temporally isolated ions were irradiated by a pulsed UV laser to detach photoelectrons. Photoelectrons were collected by the MB-PES to measure the TOF of photoelectrons. Time of flights of photoelectrons were converted to the kinetic energy of photoelectrons and the binding energy of electrons.

S. Ito, *Electronic Structures and Bonding Interaction of Structurally Defined Gold/Silver Superatoms*, Springer Theses, https://doi.org/10.1007/978-981-96-2892-6_2

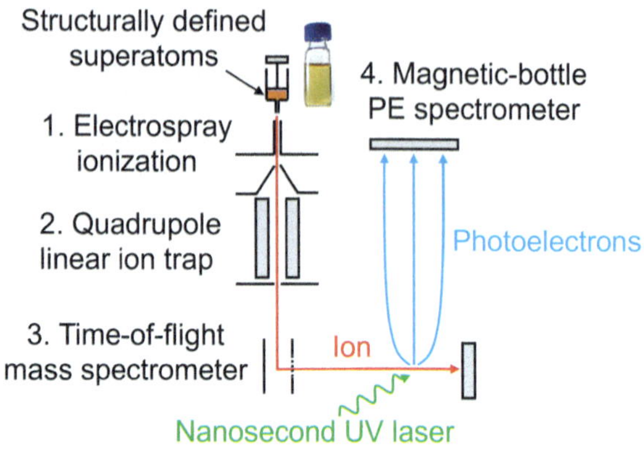

Fig. 2.1 Schematic 2D drawing of a gas-phase anion photoelectron spectrometer

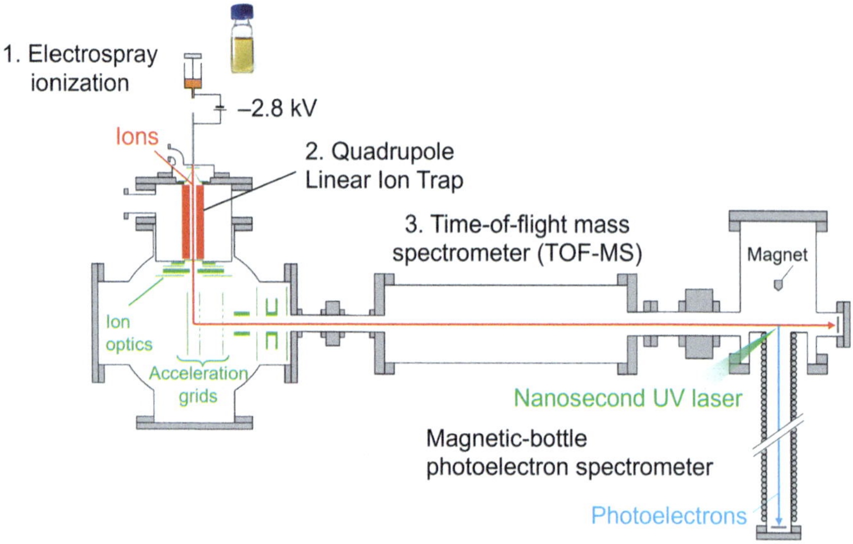

Fig. 2.2 Detail 2D drawing of a gas-phase anion photoelectron spectrometer

2.2 Electrospray Ionization Source

Mass spectrometry coupled with an ESI source was developed in 1980s by Fenn [4], a Nobel laureate in chemistry in 2002. An ESI method is known as a mild ionization method and applied to the mass spectrometry of fragile molecules and biomolecules. Since the discovery of the magic sizes of glutathione-protected gold clusters by using ESI mass spectrometry [1, 2], ESI mass spectroscopy is commonly used to determine the composition of ligand-protected gold/silver clusters.

Figure 2.4 shows the detailed drawing of the ESI source of the apparatus. A glass syringe of 1 mL was filled with a solution of ligand-protected Au/Ag clusters and

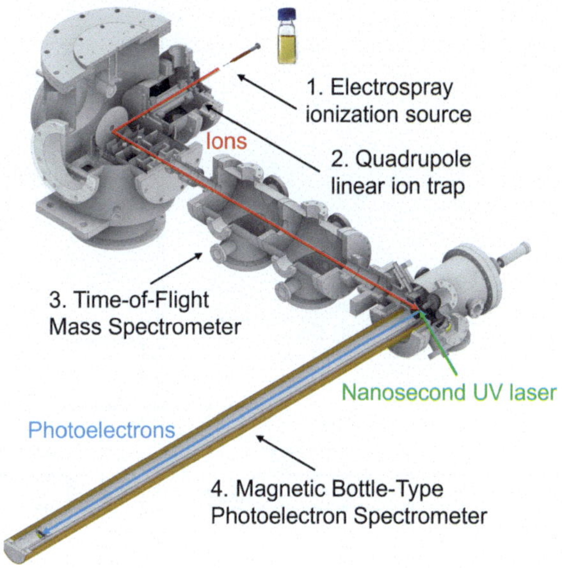

Fig. 2.3 3D drawing of a gas-phase anion photoelectron spectrometer

set on a syringe pump. The solution was pumped to a spray needle made of stainless steel (27 gauge, outer diameter = 0.41 mm, inner diameter = 0.19 mm) through a Teflon tube (inner diameter = 1/16 inch) with a typical flow rate of 3 μL/min. Typically − 3 kV was applied to the spray needle to produce anions by ESI. Charged microdroplets generated by ESI were introduced into the vacuum chamber through a desolvation capillary made of stainless steel (21 gauge, outer diameter = 0.81 mm, inner diameter = 0.51 mm) which was placed in a straight line with the spray needle. The desolvation capillary was typically heated up to 100 ℃ by a resistive heating: an electric current of 3.8 A was applied through the capillary. A ring electrode called as a focusing ring was placed at the exit of the desolvation capillary to push ions toward skimmer 1. The diameter of the hole of the focusing ring and skimmer 1 was 8 and 1.5 mm, respectively. The typical voltages applied to the focusing ring and skimmer 1 were − 80 and − 30 V, respectively. Applying a high negative voltage on the focusing ring accelerated anions, increased the collision energy of gas molecules and anions, and led to collision-induced dissociation (CID) of anions. The details of CID will be introduced in Chap. 6. The first room just after the ESI source was pumped by a rotary pump (E2M30, Edwards) and the pressure was about 200 Pa. The pumping system of the experimental setup is summarized in Table 2.1.

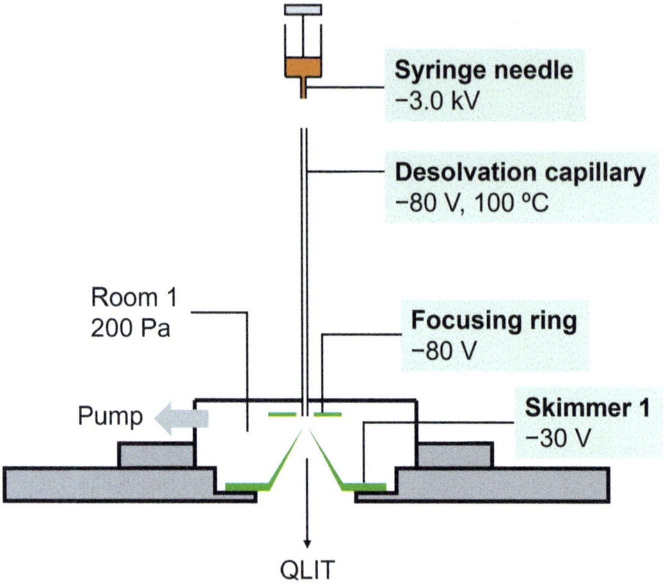

Fig. 2.4 Detail drawing of an electrospray ionization source

Table 2.1 Specification of the pumping system of the apparatus

Chamber	Rotary pump	Mechanical pump	Turbomolecular pump	Pressure (Pa)
Room 1	E2M30 (Edwards)	–	–	200
Room 2	E2M40 (Edwards)	EH500 (Edwards)	–	8
Acceleration chamber	E2M40 (Edwards)	PMB001C (ULVAC)	FW2300 (ULVAC)	1×10^{-3}
TOF tube	E2M30 (Edwards)	–	UTM-500A (ULVAC), PT500 (Mitsubishi heavy industry)	5×10^{-5}
Detector chamber	PASCAL 2021 (Alcatel)	–	PT300 (Mitsubishi heavy industry)	2×10^{-6}

2.3 Quadrupole Linear Ion Trap

2.3.1 Principle

A QLIT was developed after the invention of quadrupole mass filter by Wolfgang Paul, a Nobel laureate in chemistry in 1989 [5]. Ions were radially confined by a quadratic effective potential produced by radio frequency (RF) voltage applied on the quadrupole and axially confined by a stopping potential applied to the exit electrode. The QLIT implemented in my doctoral study is focused on the trapping of the heavy ions with m/z ~5000–10,000. Before going to the experimental setup, the theoretical background of the QLIT is overviewed [6].

Consider a particle with charge q and mass m moving in an oscillating electric field without a static field, $\boldsymbol{E}(\boldsymbol{r}) \cos \Omega t$ where $\boldsymbol{E}(\boldsymbol{r})$ and Ω is an amplitude and an angular frequency of the oscillating field. Classical equation of motion is

$$m\ddot{\boldsymbol{r}} = qE(\boldsymbol{r}) \cos \Omega t. \tag{2.1}$$

Assuming a simple case where $\boldsymbol{E}(\boldsymbol{r})$ is independent from $\boldsymbol{r}$, a special solution

$$\boldsymbol{r}(t) = \boldsymbol{r}(0) - \frac{qE}{m\Omega^2} \cos \Omega t \tag{2.2}$$

can be verified immediately by substituting Eqs. (2.2) into (2.1), and one obtains the amplitude vector $\boldsymbol{a}$:

$$\boldsymbol{a} = \frac{qE}{m\Omega^2}. \tag{2.3}$$

When $\boldsymbol{E}(\boldsymbol{r})$ is depending on $\boldsymbol{r}$, i.e. quadrupole, Eq. (2.1) can no longer be solved analytically. Therefore, an approximate solution of Eq. (2.1) will be derived to grasp some general properties of the motion of a charged particle in an oscillatory field.

Under the assumption that the field varies smoothly as a function of the coordinate $\boldsymbol{r}$ and that the frequency Ω is high enough to keep the amplitude of the oscillation $\boldsymbol{a}$ small, it can be expected that the amplitude $\boldsymbol{a}(t)$ and phase of oscillation vary slowly with time. Therefore, the solution of Eq. (2.2) can be expressed by superimposing a smooth drift term $\boldsymbol{R}_0(t)$ and a rapidly oscillating motion $\boldsymbol{R}_1(t) = -\boldsymbol{a}(t) \cos \Omega t$:

$$\boldsymbol{r}(t) = \boldsymbol{R}_0(t) - \boldsymbol{a} \cos \Omega t. \tag{2.4}$$

We presume that $\boldsymbol{E}$ is approximated by the first order Taylor approximation around $\boldsymbol{R}_0$

$$\boldsymbol{E}(\boldsymbol{R}_0 - \boldsymbol{a} \cos \Omega t) = \boldsymbol{E}(\boldsymbol{R}_0) - (\boldsymbol{a} \cdot \nabla)\boldsymbol{E}(\boldsymbol{R}_0) \cos \Omega t \tag{2.5}$$

and that fast oscillation implies $\dot{\boldsymbol{a}} \ll \Omega \boldsymbol{a}$ and $\ddot{\boldsymbol{R}}_0 \ll \Omega \dot{\boldsymbol{R}}_0$. Substituting Eqs. (2.4) and (2.5) to (2.1), and neglecting the small time derivatives, leads to an equation

$$m\ddot{\boldsymbol{R}}_0 + m\Omega^2 \boldsymbol{a}(t) \cos \Omega t = q\boldsymbol{E}(\boldsymbol{R}_0) \cos \Omega t - q[\boldsymbol{a}(t) \cdot \nabla]\boldsymbol{E}(\boldsymbol{R}_0) \cos^2 \Omega t. \tag{2.6}$$

Assuming that the amplitude $\boldsymbol{a}$ changes in time only due to the motion along $\boldsymbol{R}_0$, $\boldsymbol{a}(t)$ can be viewed as a function of $\boldsymbol{R}_0$ and thus $\boldsymbol{E}(\boldsymbol{R}_0)$ by Eq. (2.3). Under this assumption, two $\cos \Omega t$ terms in Eq. (2.6) become identical and cancel:

$$m\ddot{\boldsymbol{R}}_0 = -q\left[\frac{qE(\boldsymbol{R}_0)}{m\Omega^2} \cdot \nabla\right]\boldsymbol{E}(\boldsymbol{R}_0)\cos^2 \Omega t. \tag{2.7}$$

Considering that a vector analysis relation

$$(\boldsymbol{E} \cdot \nabla)\boldsymbol{E} = \tfrac{1}{2}\nabla E^2 - \boldsymbol{E} \times (\nabla \times \boldsymbol{E}) \tag{2.8}$$

and the rotation of $\boldsymbol{E}$ vanishes for quasistatic field conditions, it is clear that

$$(\boldsymbol{E} \cdot \nabla)\boldsymbol{E} = \tfrac{1}{2}\nabla E^2 - \boldsymbol{E} \times (\nabla \times \boldsymbol{E}) = \tfrac{1}{2}\nabla E^2. \tag{2.9}$$

Substituting Eqs. (2.9)–(2.7) and replacing $\cos^2 \Omega t$ by its time-averaged value 1/2 leads to the time-averaged oscillatory field expressed by the following equation

$$m\ddot{\boldsymbol{R}}_0 = \frac{q^2}{4m\Omega^2}\nabla E^2. \tag{2.10}$$

If a non-zero electrostatic field $\boldsymbol{E}_S(\boldsymbol{r}) = -\nabla\boldsymbol{\Phi}_S$ exists, the total average force acting on the ion can be derived from a time-independent "effective potential" V^*

$$V^*(\boldsymbol{R}_0) = \frac{q^2E^2}{4m\Omega^2} + q\boldsymbol{\Phi}_S \tag{2.11}$$

and the equation of motion describing the smooth trajectory becomes

$$m\ddot{\boldsymbol{R}}_0 = \nabla V^*(\boldsymbol{R}_0). \tag{2.12}$$

For most of the applications, this effective potential approximation allows a satisfying characterization of the properties of arbitrary field geometries.

The ideal potential of the quadrupole fulfills $\Phi(r = r_0, \phi) = \Phi_0 \cos 2\phi$ and $\Phi(r = 0, \phi) = 0$ in the cylindrical coordinate (Fig. 2.5), where r_0 is the radius of inner space made by the quadrupole and Φ_0 is the voltage applied to the quadrupole and expressed as

$$\Phi_0 = U_0 - V_0 \cos \Omega t \tag{2.13}$$

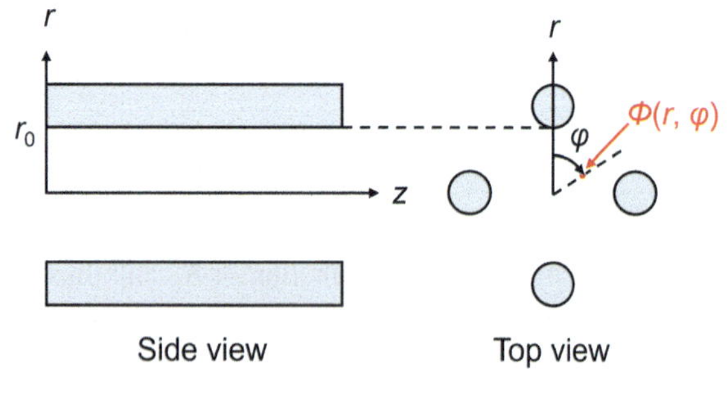

Fig. 2.5 Schematic illustration of a quadrupole

The analytical solution of the potential is described by using the reduced radius $\hat{r} = r/r_0$:

$$\Phi(r,\phi) = \Phi_0\hat{r}^2\cos 2\phi = U_0\hat{r}^2\cos 2\phi - V_0\hat{r}^2\cos 2\phi\cos\Omega t. \tag{2.14}$$

The components of the electric field $E = \left(E_r, E_\phi\right)$ are obtained as the gradient of Φ.

$$E = -\left(\frac{\partial}{\partial r}\Phi, \frac{1}{r}\frac{\partial}{\partial\phi}\Phi\right) = \frac{\Phi_0}{r_0}2\hat{r}(-\cos 2\phi, \sin 2\phi) \tag{2.15}$$

$$|\boldsymbol{E}_0| = \frac{V_0}{r_0}2\hat{r} \tag{2.16}$$

and the effective potential made by the quadrupole is

$$V^*(r,\phi) = \frac{q^2}{m\Omega^2}\frac{V_0^2}{r_0^2}\hat{r}^2 + qU_0\hat{r}^2\cos 2\phi. \tag{2.17}$$

The potential is a quadratic with respect to the reduced radius $\hat{r}$. The depth of the potential decreases with increasing the mass of ions and a RF voltage with a large amplitude or a low frequency is required to trap heavy ions. In order to produce the ideal potential in Eq. (2.14), the diameter of rods is empirically described as

$$d = 2\times 1.148r_0. \tag{2.18}$$

2.3.2 Experimental Setup

A detail drawing of the QLIT is shown in Fig. 2.6. The quadrupole consists of four stainless steel rods with the diameter of 20 mm and the length of 184 mm. The radius of an inner space surrounded by quadrupoles was 8.7 mm, which satisfies Eq. (4.18). The typical RF voltage applied to the QLIT was $V_{\mathrm{pp}} = 1000$ V, $f = 500$ kHz, and $V_{\mathrm{offset}} = -17$ V for ions with $m/z = 7400$. The voltage was decreased when m/z of ions was small. Axial confinement of ions was achieved by the stopping potential applied to skimmer 2 placed at the exit of the quadrupole. The voltage of skimmer 2 was controlled by a pulse generator to switch the trapping and the release of ions. Anions were trapped with the voltage of − 30 V for ~ 99 ms and ejected with 0 V before the acceleration of the TOF–MS by 300 μs. Timing charts of the experiment including other parts are shown in Fig. 2.7 and all timings were controlled by a delay generator (DG645, SRS). Ejected anions were focused on an acceleration region of the time-of-flight mass spectrometer by a lens 1 and lens 2 with the typical voltage of −30 and +100 V, respectively. The pressure of the chamber named as room 2 was about 8 Pa (Table 2.1).

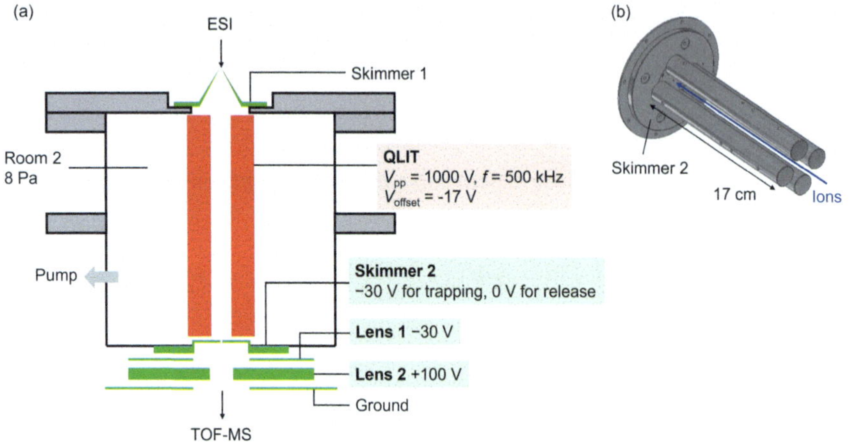

Fig. 2.6 **a** Detail drawing of a quadrupole linear ion trap. **b** 3D drawing of a quadrupole linear ion trap and skimmer 2

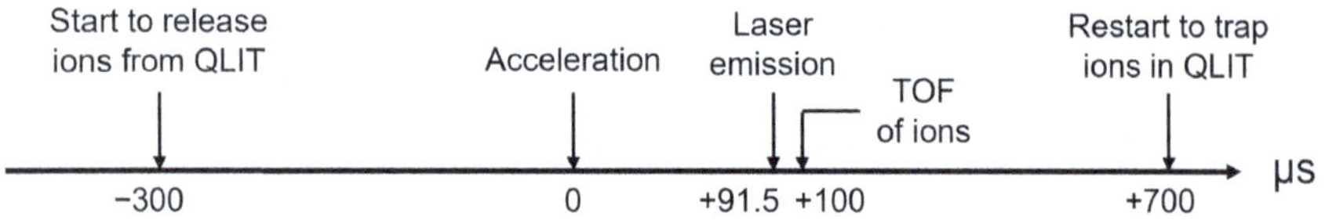

Fig. 2.7 Timing charts of the experiment

A variable-frequency RF power supply was also developed on a printed circuit board according to the previous reports [7, 8], because commercial RF power supplies are fixed to one frequency and expensive. The power supply consists of a transistor oscillator, a LC tank circuit electrically isolated from the transistor oscillator, and an automatic gain control (AGC) circuit, as shown in the circuit diagram (Fig. 2.8). Four external DC power supplies are connected to the RF power supply: DC 24 V to activate ICs and fans, DC 215 V to supply the current to the transistor oscillator (B1 in Fig. 2.8), DC 0–10 V to the AGC circuit to adjust the amplitude of the RF voltage (V4), and DC − 50–0 V to apply the DC offset on the RF voltage (between L4 and L5). The frequency of the RF output was adjusted by the capacitance of a LC tank circuit (C1) with a typical value of 700 pF. The capacitance of the QLIT was estimated to be very small (<pF) because the oscillation frequency did not change after the connection to the QLIT.

The implementation of the QLIT significantly enhanced the ion intensity and photoelectron counts of structurally defined superatoms about two orders of magnitudes. The QLIT not only contributed the quality improvement of anion PES but also allowed us to conduct advanced measurement such as the laser-fluence dependence measurement shown in Chap. 3 and anion PES of exotic species in the gas phase shown in Chap. 6.

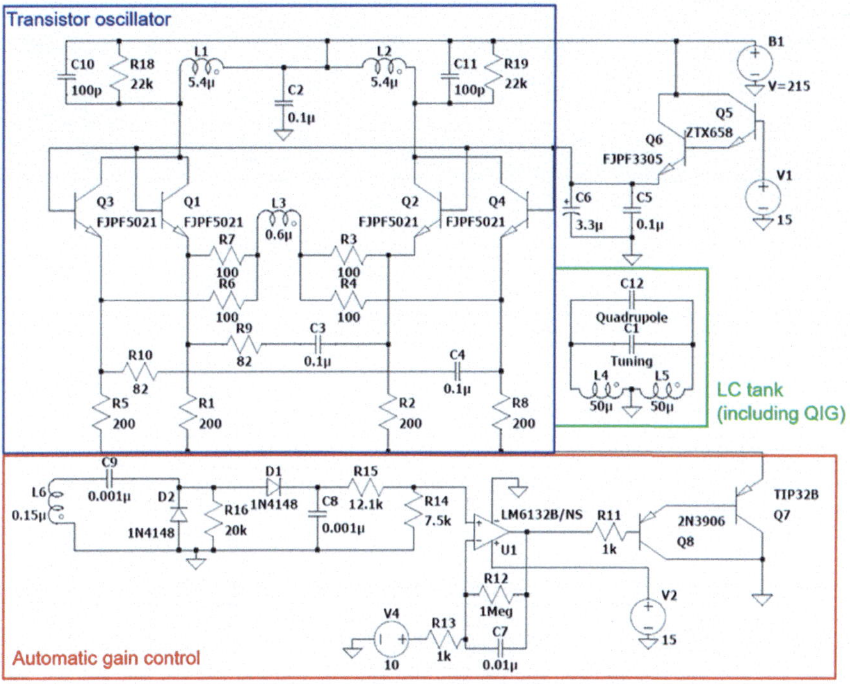

Fig. 2.8 Circuit diagram of a RF power supply. All inductances (L1 to L6) are coupled in a single air-core transistor

2.4 Time-of-Flight Mass Spectrometer

2.4.1 *Principle*

A TOF–MS is a common method to determine the mass-to-charge ratio (*m/z*) of ions in the gas phase. Ions are accelerated by a pulsed electric field and obtain the same kinetic energy by the acceleration if they have the same charge. Ions fly through the field-free time-of-flight tube toward the ion detector. Heavier ions have smaller velocities and longer time-of-flights. Schematic illustration of a typical TOF–MS is shown in Fig. 2.9.

Energy conservation between an electric energy and a kinetic energy of the ion is written as follows:

$$Vq = \frac{1}{2}mv^2 \tag{2.19}$$

where V, q, m, and v are the electric potential difference between the initial position of ions and the ground plate called as an acceleration voltage, the charge, mass, and velocity of the ion, respectively. v is constant during the field-free flight and can be

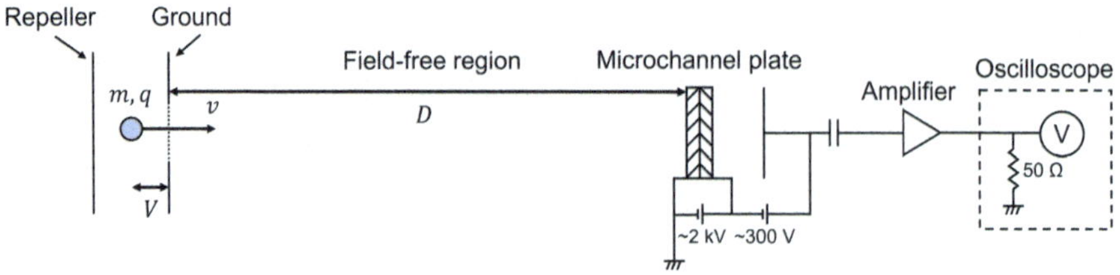

Fig. 2.9 Schematic illustration of a time-of-flight mass spectrometer for anions

determined from the known length of the flight tube (D) and the measured time of flight of ions (t). Thus,

$$v = \frac{D}{t} \tag{2.20}$$

and the mass-to-charge ratio of ions is determined from the following equation by substituting Eqs. (2.20) into (2.19):

$$\frac{m}{q} = \frac{2V}{d^2}t^2. \tag{2.21}$$

Ions are detected by a microchannel plate (MCP) where incident particles are amplified as electrons at high gain and high time resolution. The input side of the MCP is grounded while the output side is biased to the high positive voltage typically at ~2 kV. The amplified electrons are collected by the positively biased electrode and transferred to the preamplifier through the coupling capacitor which isolate the high-voltage MCP and the amplifier and the oscilloscope. The amplified ion current is converted to the voltage and monitored by the oscilloscope. The ion signals are observed as the negative voltage on the oscilloscope.

To improve the resolution of the TOF–MS, Wiley-McLaren type electrodes [9] (Fig. 2.10) are widely implemented where three electrodes are used to accelerate ions. The electric field between a repeller and an extractor is set to gentle compared to that between an extractor and a ground. The total time of flight is written as following equation using parameters in Fig. 2.10.

$$T = \frac{\sqrt{2m}}{qE_s}\sqrt{qsE_s} + \frac{\sqrt{2m}}{qE_d}\left(\sqrt{qsE_s + qdE_d} - \sqrt{qsE_s}\right) + \frac{\sqrt{2m}D}{\sqrt{qsE_s + qdE_d}} \tag{2.22}$$

The advantage of the Wiley-McLaren type TOF–MS is the space focusing along the transverse axis because an ion initially closer to the detector acquires less energy and is eventually overtaken by an ion which is less close and gets larger energy.

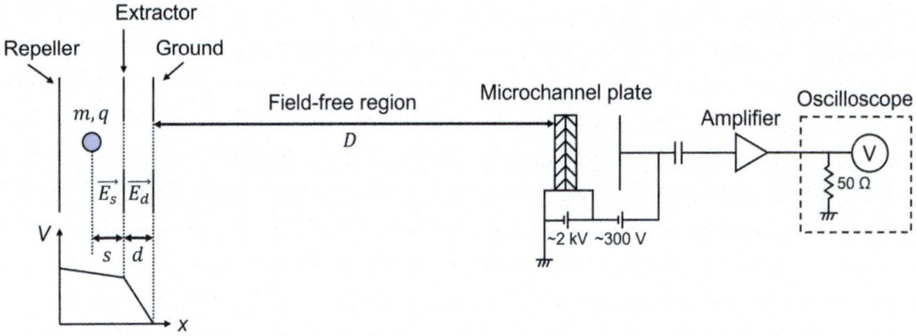

Fig. 2.10 Schematic illustration of a Wiley-McLaren type time-of-flight mass spectrometer

2.4.2 *Experimental Setup*

The actual setup is shown in Fig. 2.11. The geometrical value *s*, *d*, *D* was set to 19.5, 30.5, and 1609 mm, respectively. The acceleration voltage applied to the repeller and the extractor was set to 13.9 and 11.6 kV, respectively. In addition to the acceleration grids, a deflector and an einzel lens is implemented just after the acceleration grid to control the beam direction and to focus the ions along the radial direction. The typical voltages applied on the vertical and horizontal deflector and the einzel lens were + /− 200 V, + /− 400 V and −5.2 kV, respectively for ions with m/z = 7400. Ions were accelerated just in front of the detector by 2.4 kV to improve the detection efficiency and detected by a microsphere plate (Z033DA, El-Mul) with 2.5 kV applied. Ion currents were amplified 25 times by a preamplifier (SR445, SRS) and recorded with a 500 MHz oscilloscope (TDS520, Tektronix). The pressure of an acceleration chamber, a time-of-flight chamber, and a detection chamber was about 1×10^{-3}, 5×10^{-5}, and 2×10^{-6} Pa, respectively (Table 2.1). Under that high vacuum condition, ions does not collide with any gas molecules during TOF–MS.

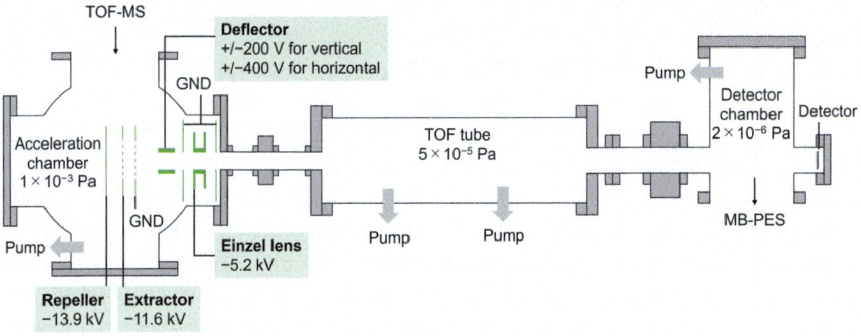

Fig. 2.11 A detailed drawing of a time-of-flight mass spectrometer

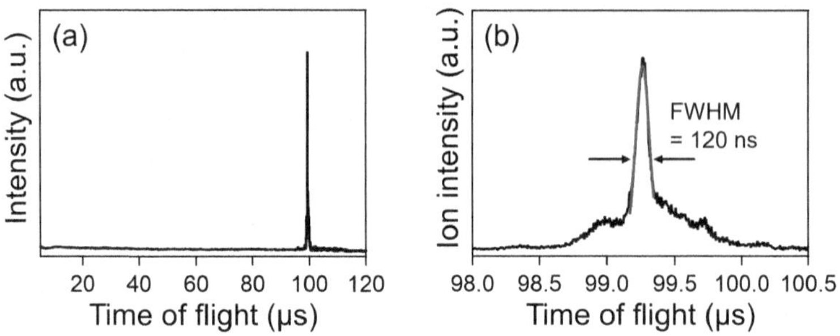

Fig. 2.12 A typical TOF mass spectrum of $[Au_{25}(SC_8H_9)_{18}]^-$

A typical mass spectrum of a representative gold cluster $[Au_{25}(SC_8H_9)_{18}]^-$ is shown in Fig. 2.12. The resolution of the TOF–MS ($m/\Delta m = t/2\Delta t$) was about 400 and enough to distinguish the difference of one Au atom ($m = 197$) within a typical cluster ($m = 7000 - 10{,}000$). TOF mass spectra were calibrated by using that of $(CsI)_nI^-$ cluster or that of structurally defined superatoms with known masses and charges by using the following equation

$$m/z = (at + b)^2 \tag{2.23}$$

where a and b is fitting parameters while m/z and t is the mass-to-charge ratio and the TOF of ions. When converting the TOF to m/z, the ion intensity should be converted simultaneously to maintain the peak area using the following equation:

$$I(t)dt = I(m)dm \tag{2.24}$$

$$I(m) = I(t)\frac{dt}{dm} = \frac{1}{2at}I(t). \tag{2.25}$$

2.5 Magnetic-Bottle Type Photoelectron Spectrometer

2.5.1 *Principle*

A MB-PES was developed by Kruit and Read to collect electrons emitted to all angles by using an inhomogeneous magnetic field [10]. The high collection efficiency is crucial to conduct anion PES as the density of ions is very low. The MB-PES is composed of a strong permanent magnet with ~1 T and a weak magnetic bottle with ~10^{-3} T made by a solenoid coil surrounding a chamber (Fig. 2.13).

Assuming an electron with an energy E and a velocity v initially emitted at an angle θ_i relative to the direction of the magnetic field around a permanent magnet

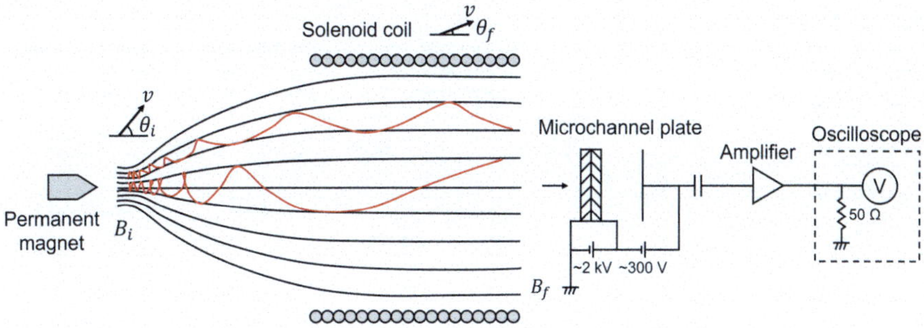

Fig. 2.13 Schematic illustration of a magnetic-bottle photoelectron spectrometer

B_i, the electron feels the central force

$$F = evB_i \sin\theta_i \tag{2.26}$$

where e is an elementary charge and undergoes a helical motion. The equation of a uniform circular motion of the electron is

$$m\frac{(v\sin\theta_i)^2}{r_i} = evB_i \sin\theta_i \mathrel{\therefore} r_i = \frac{mv\sin\theta_i}{eB_i} \tag{2.27}$$

and the angular momentum of the circular motion is

$$l_i = p_i r_i \sin\theta_i = \frac{m^2v^2\sin^2\theta_i}{eB_i}. \tag{2.28}$$

The same circular motion is also observed in the magnetic bottle with B_f. If the variation of the magnetic field with the axis of the magnetic bottle is adiabatic, the angular momentum is a conserved quantity.

$$l_i = l_f \mathrel{\therefore} \frac{\sin\theta_f}{\sin\theta_i} = \sqrt{\frac{B_f}{B_i}} \tag{2.29}$$

Considering that the ratio of B_f and B_i is about 1000, θ_f is 1.8° even if $\theta_i = 90°$ (perpendicular to the axis of the magnetic bottle) and all electrons are almost parallelized by the magnetic bottle. The velocity along the axis of the magnetic bottle is increased from $v\cos\theta_i$ to

$$v\cos\theta_f = v\left(1 - \sin^2\theta_f\right) = v\left(1 - \sin^2\theta_i\frac{B_f}{B_i}\right) \tag{2.30}$$

and this is approximately equal to v. Although the time of flight during the parallelization depends on the initial direction of the electron, theoretical simulation [10] showed that the contribution of the time is enough small to provide an appropriate resolution.

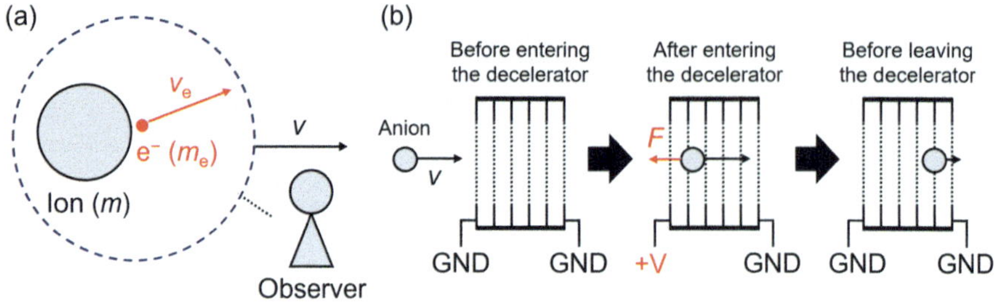

Fig. 2.14 Schematic illustration of **a** a Doppler effect and **b** a momentum decelerator in photoelectron spectroscopy

In anion PES, photoelectrons were detached from moving ions during TOF–MS. Under this situation, the velocity of ions in a laboratory coordinate system is the sum of the velocity of photoelectrons and that of ions (Fig. 2.14a), called as the Doppler effect. The Doppler effect results in a decrease of the resolution. The minimum and maximum velocity of ions in a laboratory coordinate system $v_{e,\text{lab}}$ is

$$v_{e,\text{lab}} = v_e \pm v \quad (2.31)$$

where $v_{e,\text{lab}}$ and v is the ideal velocity of the photoelectron and the ion, respectively. The energy of the photoelectron is

$$E_{e,\text{lab}} = \frac{1}{2}m_e v_{e,\text{lab}}^2 = E_e + \frac{m_e}{m}E \pm 2\sqrt{\frac{m_e}{m}E_e E} \quad (2.32)$$

where E_e, m_e, E, m is the energy and mass of the electron and the ion, respectively. Therefore, the difference of the energy between the fastest and slowest photoelectrons is $4\sqrt{\frac{m_e}{m}E_e E}$.

The Doppler effect can be compensated by the deceleration of ions before PES. Ion deceleration was achieved by a momentum decelerator (Fig. 2.14b) [11]. The momentum decelerator is composed of the stacked ring electrodes with a mesh to produce the homogeneous electric fields in it. Before anions enter the decelerator, both the entrance and the exit are grounded. After anions enter it, a positive high voltage pulse is applied to the entrance to pull back anions. The high voltage is pulsed to ground before ions leave the decelerator. All anions experience the same decelerating force for the same period and thus lose the same amount of momentum. The momentum deceleration reduces the kinetic energy distribution after the deceleration compared to the energy deceleration by the static potential [11].

2.5.2 Experimental Setup

A detailed drawing of the MB-PES is shown in Fig. 2.15. Photoelectrons were detached by irradiating the nanosecond UV laser on the flying ions. The laser used in this study was a third, fourth, and fifth harmonics of Nd:YAG laser and an ArF excimer laser. The specification of the laser is listed in Table 2.2. The laser timing was roughly estimated by multiplying the TOF of ions by 0.915 and controlled by a delay generator (Fig. 2.7). The actual timing of the laser emission was monitored by a photodiode (S12271, Hamamatsu Photonics) and used as a zero point of the TOF of photoelectrons.

The magnetic bottle was composed of a permanent magnet with a cone and a solenoid coil surrounding the TOF tube of photoelectrons with a length of 2 m. Electric current of 2.8 A was passed through the coil to produce a weak magnetic field. Photoelectrons were accelerated just in front of the detector by 580 V to improve the detection efficiency and detected by a microchannel plate with 1.8 kV applied and the photoelectron current was amplified 5 times by a preamplifier (SR445, SRS). The

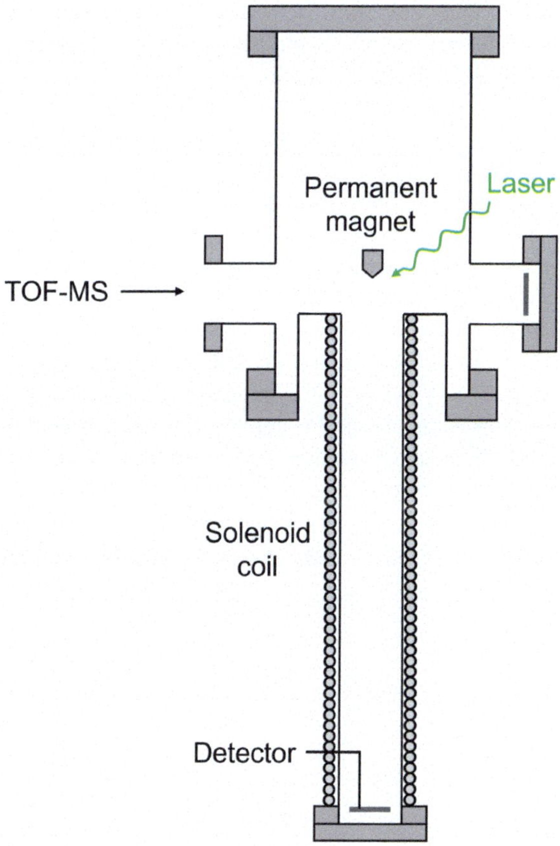

Fig. 2.15 Detail drawing of a magnetic-bottle type photoelectron spectrometer

Table 2.2 Specification of the laser used in this study

Laser	Manufacturer	Wavelength (nm)	Photon energy (eV)	Frequency (Hz)
GCR-100	Spectra physics	355	3.49	10
		266	4.66	10
LS-2134UT	LOTIS TII	213	5.82	20
PSX-100	MPB communications	193	6.42	100

TOF of photoelectrons was recorded by a multichannel scaler (SR430, SRS) with the discriminating level of -10 mV.

PE spectra were calibrated by using those of iodide ions with known electron binding energies $\left(E_{\mathrm{b}}\left({}^2\mathrm{P}_{1/2}\right) = 3.059\ \mathrm{eV}, E_{\mathrm{b}}\left({}^2\mathrm{P}_{3/2}\right) = 4.002\ \mathrm{eV}\right)$ by using the following equation

$$E = \left(\frac{3.22}{t+a}\right)^2 + b \tag{2.33}$$

where a and b is fitting parameters while E and t is the kinetic energy and the TOF of photoelectrons. In the conversion from the TOF to the energy of electrons, photoelectron counts should be converted simultaneously to maintain the peak area by using the following equation:

$$I(t)dt = I(E)dE \therefore I(E) = I(t)\frac{dt}{dE} \propto t^3 I(t). \tag{2.34}$$

A typical PE spectra of iodide ion and the corresponding calibration curve are shown in Fig. 2.16. Iodide ions were accelerated to 640 eV and decelerated by a momentum decelerator by applying 1.44 kV for 1 μs. Two sharp peaks were assigned to the transition to the neutral state ${}^2\mathrm{P}_{1/2}$ and ${}^2\mathrm{P}_{3/2}$, respectively. The full width at half maximum of the peak was 23 meV with $E_{\mathrm{e}} = 0.43$ eV and 85 meV with $E_{\mathrm{e}} =$ 2.8 eV. In PES of structurally defined superatoms, ions were typically accelerated up to 13 keV while the momentum deceleration was not conducted to retain the high ion intensity and PE counts. Under the condition with $E = 13$ keV and $m =$ 7400, the peak broadening due to the Doppler effect is expected to be ~110 $\sqrt{E_e}$ meV. Experimental peak width at that condition was 200 meV and thus the peak width of structurally defined superatoms was governed by other factors such as the temperature effect, Franck–Condon principle, and the structural fluctuation of the ligand layer.

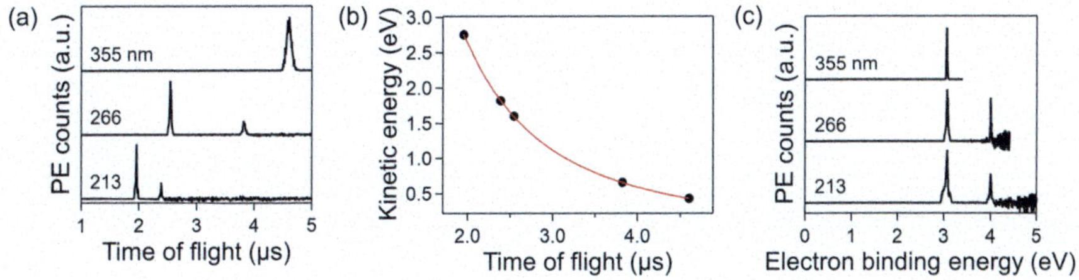

Fig. 2.16 **a** Photoelectron spectra of iodide ions plotted against time of flights. **b** Calibration curve of photoelectron spectrometer. **c** Photoelectron spectra of iodide ions plotted against electron binding energy

References

1. Negishi Y, Takasugi Y, Sato S, Yao H, Kimura K, Tsukuda T (2004) J Am Chem Soc 126:6518
2. Negishi Y, Nobusada K, Tsukuda T (2005) J Am Chem Soc 127:5261
3. Hirata K, Yamashita K, Muramatsu S, Takano S, Ohshimo K, Azuma T, Nakanishi R, Nagata T, Yamazoe S, Koyasu K et al (2017) Nanoscale 9:13409
4. Fenn JB, Mann M, Meng CK, Wong SF, Whitehouse CM (1989) Science 246:64
5. Paul W (1990) Rev Mod Phys 62:531
6. Gerlich D, Inhomogeneous RF fields: a versatile tool for the study of processes with slow ions in advances in chemical physics. Wiley, New York, pp 1–176
7. O'Connor PB, Costello CE, Earle WE (2002) J Am Soc Mass Spectrom 13:1370
8. Mathur R, O'Connor PB (2006) Rev Sci Instrum 77:114101
9. Wiley WC, McLaren IH (1955) Rev Sci Instrum 26:1150
10. Kruit P, Read FH (1983) J Phys E 16:313
11. Wang L-S, Cheng H-S, Fan J (1995) J Chem Phys 102:9480

Chapter 3
Effect of Heterometal Doping on Electron Affinities of Eight Electron Superatoms

Abstract The third chapter describes the doping effect on the eight-electron superatoms based on anion photoelectron spectroscopy (PES) of a representative gold superatom and heterometal-doped silver-based superatoms. The electron binding energies of the representative superatom $[Au_{25}(PET)_{18}]^-$ (PET = 2-PhC_2H_4S) were reexamined by anion PES on a significantly intensified mass-selected ion beam. Laser fluence-dependent PE spectra and pump–probe PES revealed that the previous PE spectra were contaminated by PE signals due to the two-photon electron detachment via long-lived photoexcited states. Anion PES on silver-based superatoms $[MAg_{24}(DMBT)_{18}]^-$ (M = Ag, Au), $[MAg_{24}(DMBT)_{18}]^{2-}$ (M = Pd, Pt) having icosahedral MAg_{12} cores with 8 electrons showed that the electron affinity of the neutral Ag_{13} superatom (~4 eV) was not appreciably affected by doping Au atom at the center but was reduced by ~2 eV by doping Pd or Pt. The doping effect can be rationalized by the charging effect and the two-step jellium potential.

Keywords Superatom · Gold cluster · Silver cluster · Doping · Two-photon photoelectron detachment

3.1 Introduction

Ligand-protected gold/silver clusters with precisely defined compositions and structures constitute a growing class of nanomaterials with unique physicochemical properties [1–10]. Their intrinsic stabilities and novel properties are intimately related to the electronic structures, which are sensitive to geometric structures and rationalized by the superatomic concept [1, 11, 12]. In this model, the valence electrons are accommodated in quantized superatomic orbitals (1S, 1P, 1D, 2S, …) distributed over the metallic cores and the protected clusters gain a special stability when the electronic shells are closed. Thus, it is important to understand how the electronic structures can be manipulated by changing their structural parameters such as the size, the geometric structures and composition of the cores, the total charge of the

S. Ito, *Electronic Structures and Bonding Interaction of Structurally Defined Gold/Silver Superatoms*, Springer Theses, https://doi.org/10.1007/978-981-96-2892-6_3

clusters, and the nature of the ligands for the rational development of functional clusters.

A promising approach to establish the design principles of electronic structures is gas-phase photoelectron spectroscopy (PES) [13–19] on ligand-protected Au/Ag cluster anions whose geometric structures were determined by X-ray crystallography. Since 2017 [13], a PES apparatus introduced in Chap. 2 without an ion trap has been applied to representative ligand-protected Au/Ag superatoms with 8 electrons, including $[Au_{25}(PET)_{18}]^-$ (PET = 2-$C_6H_5C_2H_4S$; referred to as $\mathbf{Au_{13}}^-$) [20, 21] and $[XAg_{24}(DMBT)_{18}]^z$ (DMBT = 2,4-$(CH_3)_2C_6H_3S$; $z = -1$ for X = Ag and Au and $z = -2$ for Pd and Pt; referred to as $\mathbf{Ag_{13}}^-$, $\mathbf{AuAg_{12}}^-$, $\mathbf{PdAg_{12}}^{2-}$, and $\mathbf{PtAg_{12}}^{2-}$) [22–24]. The crystal structures of these clusters are divided into the superatomic cores and surrounding ligand shells, as shown in Table 3.1. All clusters have icosahedral M_{13} superatomic cores with a closed electron configuration $(1S)^2(1P)^6$ and can be viewed as structurally defined superatoms. Thus, these superatoms are an ideal platform to investigate the doping effect on the electronic structure of structurally defined superatoms. Using PES for these 8e superatoms, the effect of the composition of the cores and the total charge of the clusters on adiabatic electron affinities (*AEA*s) of the corresponding 7e superatoms has been studied (Table 3.1). The AEAs of $\mathbf{Au_{13}}^0$, $\mathbf{Ag_{13}}^0$, and $\mathbf{AuAg_{12}}^0$ were determined to be 2.36, 2.02, and 2.08 eV, respectively, indicating that the electronic structure of icosahedral M_{13} does not depend so much on whether M is Au or Ag [14, 15]. In contrast, the AEAs of the Pd/Pt-doped anions $\mathbf{PdAg_{12}}^-$ and $\mathbf{PtAg_{12}}^-$ (0.61 and 0.60 eV, respectively) were significantly smaller than that of undoped $\mathbf{Ag_{13}}^0$ (2.02 eV) [15]. This doping effect was explained by the reduction of the formal charge in the core potential upon the replacement of Ag^+ with X^0 (X = Pd and Pt) and/or the upward shift of the apparent vacuum level by the presence of a repulsive Coulomb barrier (RCB) [25, 26] between the detached electron and remaining anions $\mathbf{PdAg_{12}}^-$ and $\mathbf{PtAg_{12}}^-$. Table 3.1 also lists the calculated AEA values of these clusters. All the reported experimental AEA values were significantly (~0.6–1.4 eV) smaller than the calculated AEAs [15, 27, 28].

In the previous studies [13–15, 18], a high-fluence laser had to be used when measuring the PE spectra of structurally defined superatoms because of the small beam intensities of the target anions: the duty cycle of pulsed extraction from the continuous electrospray ionization (ESI) beam to the time-of-flight mass spectrometer (TOF–MS) was only ~0.2%. Such a setup forced the long measurement for several hours to take one PE spectrum and prevented us from studying the laser fluence dependence of the PE signals to avoid multiphoton processes. To remedy this situation, I installed the quadrupole linear ion trap (QLIT) as introduced in Chap. 2 as an interface between the continuous ESI source and the pulsed TOF–MS. As a result, the ion intensity was increased by about two orders of magnitude. The purpose of this study was to remeasure the PE spectra of structurally defined superatoms using the updated apparatus and to discuss about the effect of heterometal doping on the electronic structure of superatomic orbitals. It was found that the previously reported AEAs were underestimated due to the involvement of PE detachment from photoex-

Table 3.1 Structurally defined superatoms used in this study

Cluster	$[Au_{25}(PET)_{18}]^-$ (Au_{13}^-)	$[XAg_{24}(DMBT)_{18}]^{2-}$			
		X = Ag^+ (Ag_{13}^-)	Au^+ ($AuAg_{12}^-$)	Pd^0 ($PdAg_{12}^{2-}$)	Pt^0 ($PtAg_{12}^{2-}$)
Ligand shell	6 × $Au_2(PET)_3$	6 × $Ag_2(DMBT)_3$			
Superatomic core	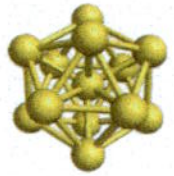				
	Au@Au_{12}[a]	X@Ag_{12}[b]			
AEA_{exp} (eV)[c]	2.36[d]	2.02[e]	2.08[e]	0.61[e]	0.60[e]
AEA_{cal} (eV)[f]	2.94[g] / 3.23[h]	3.46[e, i]	3.51[e, i]	1.17[e, i]	1.22[e, üi]
AEA_{exp} (eV)[j]	3.45	3.97	4.00	2.00	2.08

[a]References [20, 21]. [b]References [22–24]. [c]Experimental AEAs of the ionized forms determined previously. [d]Reference [14]. [e]References [15]. [f]Theoretical AEAs of the oxidized forms reported previously. [g]Reference [27]. [h]Reference [28]. [i]Ligands were simplified as SCH_3. [j]Experimental *AEA*s of the corresponding oxidized forms determined in the present study

cited states. The AEAs were redetermined from the PE spectra recorded at a smaller laser fluence to suppress the multiphoton processes (Table 3.1) and the effect of the heterometal doping was discussed.

3.2 Method

3.2.1 General

All solvents except for tetrahydrofuran, hydrogen tetrachloroaurate tetrahydrate ($HAuCl_4 \cdot 4H_2O$), sodium borohydride ($NaBH_4$), silver nitrate ($AgNO_3$), (triphenylphosphine)gold(I) chloride ($AuClPPh_3$), and triethylamine were purchased from FUJIFILM Wako Pure Chemical Corporation. Phenylethanethiol (PET-H), 2,4-dimethylbenzenethiol (DMBT-H), tetraphenylphosphonium bromide (PPh_4Br), tetraphenylphosphonium chloride (PPh_4Cl), and 3,5-bis(trifluoromethyl)ethynylbenzene ($Ar^FC{\equiv}CH$) were purchased from Tokyo Chemical Industry. Tetrahedrofuran was purchased from Kanto Chemicals. Tetraoctylammonium bromide (TOABr), palladium acetate ($Pd(OCOCH_3)_2$), and bis(triphenylphosphine)platinum chloride ($PtCl_2(PPh_3)_2$) were purchased from Sigma-Aldrich. Water used in this experiment was Milli-Q grade (>18 MΩ·cm).

UV–Visible absorption spectra were recorded by a Jasco V-630 spectrometer using dichloromethane as the solvent. Electrospray ionization mass spectrometry (ESI–MS) was performed using a Bruker compact ESI-Q-TOF mass spectrometer. Obtained mass spectra were calibrated using $[(CsI)_nI]^-$ as an external reference.

3.2.2 Synthesis and Characterization

$[Au_{25}(PET)_{18}]^-(TOA^+)$ ($\mathbf{Au_{13}}^-(TOA^+)$)

Synthesis was conducted based on the previous report [29]. $HAuCl_4 \cdot 4H_2O$ (262.3 mg, 635 μmol) and TOABr (387.5 mg, 715 μL) were dissolved in 35 mL of THF in a 100 mL eggplant flask and stirred for 10 min. PET-H (852 μL, 6.35 mmol) was added to the orange solution and stirred until the solution became colorless. Then, a freshly prepared ice-cold aqueous $NaBH_4$ solution (243.7 mg in 10 mL of water) was added all at once. The dark brown solution was continuously stirred for 2 days. The product was filtered to remove precipitates and the solvent was removed by rotary evaporation. An excess amount of MeOH was added to the dark oily product and the precipitate was collected by centrifugation. The precipitate was washed with MeOH and extracted with dichloromethane. The extract was dried by rotary evaporation and in vacuo and $[Au_{25}(PET)_{18}]^-(TOA^+)$ was obtained as a brown solid. The yield was 44.8 mg.

$[Ag_{25}(DMBT)_{18}]^-(PPh_4^+)$ (**$Ag_{13}^-(PPh_4^+)$**)

Synthesis was conducted based on the previous report with modifications [22]. $AgNO_3$ (38.7 mg, 228 μmol) was dissolved in MeOH (2 mL) with the aid of sonication and then was added to DCM (18 mL). The solution was stirred at room temperature and then added to DMBT-H (120 μL, 894 μmol) and the colorless solution turned into a yellow suspension. PPh_4Br (5.9 mg, 14 μmol) dissolved in MeOH (500 mL) was added to the suspension and cooled in an ice bath for 30 min. Then, a freshly prepared ice-cold aqueous $NaBH_4$ solution (14.6 mg in 1 mL of water) was added dropwise to the suspension. The color of the suspension gradually changed from yellow to brown. The reaction was kept at 0 °C for 6 h and then left to stand in a freezer (–20 °C) for 18 h. The product was filtered to remove the dark yellow solid and the solvent was removed by rotary evaporation. The oily product was washed with hexane and extracted with a 1:1 mixture of MeOH and DCM. The extract was dried by rotary evaporation and extracted with DCM. The extract was dried by rotary evaporation and in vacuo to afford $[Ag_{25}(DMBT)_{18}]^-(PPh_4^+)$. The yield was 6.5 mg.

$[AuAg_{24}(DMBT)_{18}]^-(PPh_4^+)$ (**$AuAg_{12}^-(PPh_4^+)$**)

Synthesis was conducted based on the previous report with modifications [24]. A 30 mL glass vial was charged with 37.4 mg of $AgNO_3$ (0.220 mmol) dissolved in 2 mL of MeOH with the aid of sonication. Stirring (280 rpm) was started in an ice bath, and 1.5 mg of $AuClPPh_3$ (0.003 mmol) dissolved in 18 mL of ice-cold DCM was added. The solution was cooled in an ice bath and stirred continuously for 5 min. Then, 110 μL of DMBT-H (0.82 mmol) was added to the vial, yielding a yellow precipitate. The suspension was cooled in an ice bath for 20 min under constant stirring. Then, 5.9 mg of PPh_4Br (0.014 mmol) dissolved in 0.5 mL of ice-cold MeOH was added to the suspension. The suspension was cooled in an ice bath (0 °C) for 30 min under constant stirring. Next, a freshly prepared $NaBH_4$ solution (15.1 mg; 0.400 mmol in 0.5 mL of ice-cold water) was added dropwise to the suspension over a period of 10 min under vigorous stirring (1200 rpm). The color of the solution turned to dark gray. The mixture was stirred in an ice bath for 1 h and then kept at room temperature for ~12 h under constant stirring.

The DCM layer was separated from the aqueous layer and then filtered to remove the insoluble black precipitate. The filtrate was concentrated to 5 mL by rotary evaporation and then was added dropwise to 40 mL of MeOH. After 16 h of stirring (1200 rpm) at room temperature, a dark brown precipitate was formed. The precipitate was collected by filtration and washed three times with MeOH. Then, the precipitate was dissolved in 40 mL of DCM and was filtered to remove the yellow precipitate. The filtrate was concentrated to 5 mL and added dropwise to 40 mL of MeOH. After 5 min of stirring, the solution was filtered and the filtrate was evaporated. The dried sample was again dissolved in 5 mL of DCM and added dropwise to 40 mL of hexane. The brown precipitate that formed after 5 min of stirring was collected by filtration and washed three times with hexane. The precipitate was dissolved in 10 mL of DCM, layered with 20 mL of hexane, and then left in a freezer (5 °C). Green crystals were formed after one week.

$[PdAg_{24}(DMBT)_{18}]^{2-}(PPh_4^+)_2$ (**$PdAg_{12}^{2-}$**$(PPh_4^+)_2$)

Synthesis was conducted based on the previous report with modifications [23]. A 30 mL glass vial was charged with 37.9 mg of $AgNO_3$ (0.252 mmol) dissolved in 2 mL of MeOH with the aid of sonication. Stirring (280 rpm) was started in an ice bath, and 4.6 mg of $Pd(OCOCH_3)_2$ (0.020 mmol) dissolved in 9 mL of DCM was added. The vial used for the dissolution of $Pd(OCOCH_3)_2$ was washed twice with 4.5 mL of DCM and the washing liquid was added to the reaction mixture. Then, 110 μL of DMBH-H (0.81 mmol) was added to the vial, yielding a yellow precipitate. The suspension was kept in an ice bath for 20 min under constant stirring. Then, 7.0 mg of PPh_4Br (0.017 mmol) dissolved in 0.5 mL of MeOH was added to the solution. The suspension was cooled in an ice bath for 30 min. Next, a freshly prepared $NaBH_4$ solution (16.7 mg; 0.441 mmol in 5 mL of ice-cold water) was added dropwise to the suspension over a period of 10 min under vigorous stirring (1200 rpm). The color of the solution turned to dark gray. After the mixture was stirred in an ice bath for 6 h, it was left to stand in a freezer (–20 °C). The purification process was the same as that for $[AuAg_{24}(SPhMe_2)_{18}]^-(PPh_4^+)$. The crude product was dissolved in 10 mL of DCM, layered with 20 mL of hexane, and then left in a freezer (5 °C). Dark brown crystals were formed after one week.

$[PtAg_{24}(DMBT)_{18}]^{2-}(PPh_4^+)_2$ (**$PtAg_{12}^{2-}$**$(PPh_4^+)_2$)

Synthesis was conducted based on the previous report with modifications [23]. A 30 mL glass vial was charged with 38.2 mg of $AgNO_3$ (0.225 mmol) dissolved in 2 mL of MeOH with the aid of sonication. Stirring (280 rpm) was started in an ice bath, and 7.11 mg of $PtCl_2(PPh_3)_2$ (0.009 mmol) dissolved in 9 mL of DCM was added. The vial used for the dissolution of $PtCl_2(PPh_3)_2$ was washed twice with 4.5 mL of DCM and the washing liquid was added to the reaction mixture. Then, 110 μL of DMBT-H (0.81 mmol) was added to the vial, yielding a yellow precipitate. The suspension was kept in an ice bath for 20 min under constant stirring. Then, 7.1 mg of PPh_4Br (0.017 mmol) dissolved in 0.5 mL of MeOH was added to the solution. The suspension was cooled in an ice bath for 30 min. Next, a freshly prepared $NaBH_4$ solution (17.8 mg; 0.471 mmol in 5 mL of ice-cold water) was added dropwise to the suspension over a period of 10 min under vigorous stirring (1200 rpm). The color of the solution turned to dark gray. After the mixture was stirred in an ice bath for 6 h, it was left to stand in a freezer (–20 °C). The purification process was the same as that for $[AuAg_{24}(SPhMe_2)_{18}]^-(PPh_4^+)$. The crude product was dissolved in 10 mL of DCM, layered with 20 mL of hexane, and then left in a freezer (5 °C). Green crystals were formed after one week.

Characterization

The successful synthesis was confirmed by electrospray ionization (ESI) mass spectroscopy (Fig. 3.1a) and UV–visible absorption spectroscopy (Fig. 3.1b). ESI mass spectra showed the well-resolved isotope pattern which matched well with the simulated one (insets of Fig. 3.1a). UV–visible absorption spectra exhibited characteristic features as reported previously [20–24]. Arrows in UV–visible absorption spectra

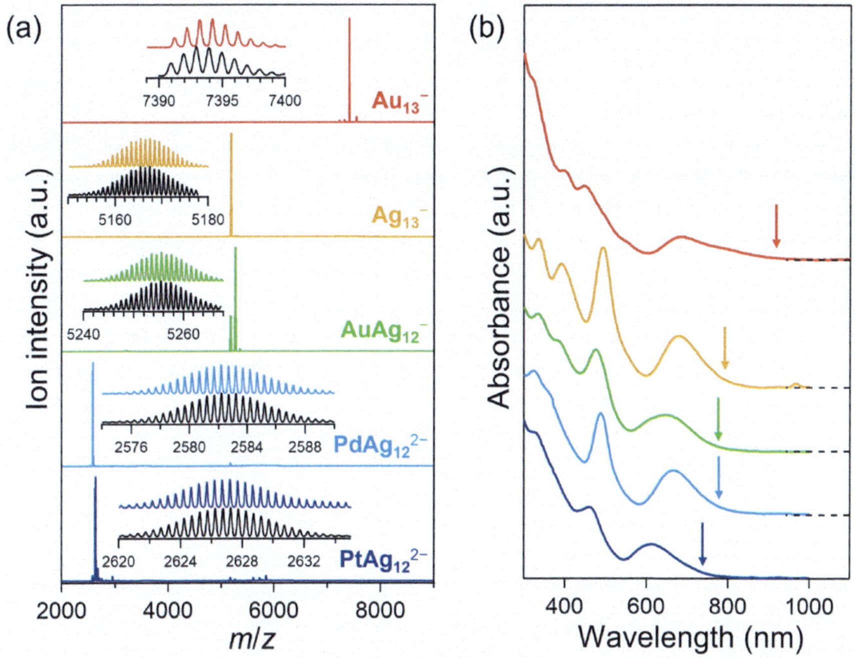

Fig. 3.1 **a** ESI mass spectra and **b** UV–visible absorption spectra of $\mathbf{Au_{13}}^-$, $\mathbf{Ag_{13}}^-$, $\mathbf{AuAg_{12}}^-$, $\mathbf{PdAg_{12}}^{2-}$ and $\mathbf{PtAg_{12}}^{2-}$. Insets in (**a**) compare the observed isotope pattern of the strongest peak (colored) and the simulated isotope pattern of each cluster (black traces). Arrows in (**b**) indicate the spectral onset

indicate the spectral onset corresponding to the optical highest occupied molecular orbital (HOMO)-lowest unoccupied molecular orbital (LUMO) gap (E_{HL}), which will be referred in the discussion later (Table 3.3).

3.2.3 Photoelectron Spectroscopy

Anion PES was conducted using a home-built apparatus equipped with an ESI source, a time-of-flight mass spectrometer (TOF–MS), and a magnetic-bottle type photoelectron spectrometer (MB-PES) as introduced in Chap. 2. Solutions of structurally defined superatoms were introduced into the apparatus via the ESI source and desolvated by a heated capillary. Ions were transferred to a quadrupole linear ion trap (QLIT) through a skimmer (1 mm diameter). Continuously flowing ions were accumulated in the QLIT for 98 ms and released 300 μs before the acceleration of the TOF–MS. The released ions were accelerated by a set of Wiley–McLaren type electrodes and were temporally and spatially separated by the TOF–MS with a mass resolution ($m/\Delta m$) of ~600. Pulsed third or fourth harmonics of an Nd:YAG nanosecond laser (Quanta-Ray GCR130, Spectra-Physics, U.S.) or pulsed ArF excimer laser (PSX-100, MPB Communications Inc.) clipped by a pinhole (1000 μm for Nd:YAG laser; 300 μm for ArF excimer laser) were irradiated to the mass-selected ion beam during the TOF–MS. The emitted photoelectrons were collected by an inhomogeneous magnetic field generated by a permanent magnet and a coil. The kinetic energies of the photoelectrons were determined by the TOF of the electrons and then converted to the electron binding energies (EBEs). The EBEs were calibrated with the PE spectrum of I^- produced by the dissociative electron attachment to CH_3I. The energy resolution was about 110 mV for an electron with a kinetic energy of 1 eV. The obtained PE spectra were smoothed by averaging the data points within ±50 meV. The adiabatic electron affinity was determined by the energy of the spectral onset where the tangential line at the inflection point intersected the X-axis. Laser fluences were measured in front of the laser head.

The dependence of the PE count on the laser fluences was measured by the same setup. The laser fluences were adjusted by inserting ND filters with optical densities of 0.1–2.0 after the clipping pinhole.

Pump-probe PES with two nanosecond laser pulses was conducted by using the same spectrometer. A pump pulse of 532 nm was generated by another Nd: YAG laser (Quanta-Ray INDI 40–20, Spectra-Physics, U.S.) synchronized by a delay generator and irradiated to ions without any pinholes. The laser timing was roughly monitored by a photodiode sensor. The ion intensity was simultaneously measured in both experiments to confirm that the change of PE counts are not due to the change of the ion intensity. Maximum time delay was limited to 160 ns considering the radius of the 532 nm laser (~2.8 mm) and the velocity of the ions (~17 mm μs^{-1}).

Photodissociation mass spectrometry (PD-MS) was conducted by using the same mass spectrometer. A nanosecond laser pulse at 266 nm was irradiated to ions at acceleration grids 1 μs before the acceleration.

3.2.4 Theoretical Calculation

Density functional theory (DFT) calculation was conducted to assign the PE spectrum of $\mathbf{Au_{13}}^-$ by using Gaussian 16 program [30]. Structural optimization was conducted by using the crystal structure of $\mathbf{Au_{13}}^{-1}$ [20, 21] as an initial structure. The calculation level was set to B3LYP functional and 6–31G(d) basis set for C, H, and S and LanL2DZ basis set for Au. Frequency calculation did not show any imaginary frequencies, indicating that the optimized structure is located at the local minimum. The vertical detachment energy of the optimized anion was calculated by subtracting the total energy of the neutral at the fixed structure of the optimized anion from that of the optimized anion. The simulated PE spectrum was constructed as follows. Firstly, all Kohn–Sham energy levels are shifted equally by the energy difference between the calculated vertical detachment energy and the energy of HOMO level. Then, each energy level was replaced by a Gaussian function with the same height and the same width. Finally, the simulated PE spectrum was constructed by summing up all the Gaussian functions.

3.2.5 Laser Fluence Dependence of the Photoelectron Detachment Process

We assume that the following three elementary steps compete in the photoelectron detachment of M^-:

$$M^- + h\nu \rightarrow M^0 + e^-(1) \tag{3.1}$$

$$M^- + h\nu \rightarrow M^{-*} \tag{3.2}$$

$$M^{-*} + h\nu \rightarrow M^0 + e^-(2) \tag{3.3}$$

where reactions (3.1), (3.2), and (3.3) correspond to direct electron detachment from M^-, photoabsorption by M^-, and electron detachment from M^{-*} in the photoexcited state. $e^-(1)$ and $e^-(2)$ indicate the electrons generated by the one-photon and two-photon process, respectively. The differential rate equations are written as follows:

$$\frac{d[M^-]}{dt} = -(k_1 + k_2)I[M^-] \tag{3.4}$$

$$\frac{d[M^{-*}]}{dt} = k_2 I[M^-] - k_3 I[M^{-*}] \tag{3.5}$$

$$\frac{d[e^-(1)]}{dt} = k_1 I[M^-] \tag{3.6}$$

$$\frac{d[e^-(2)]}{dt} = k_3 I[M^{-*}] \tag{3.7}$$

where k_1, k_2, and k_3 are the rate constants for reactions (3.1), (3.2), and (3.3), respectively, and I is the photon density. Equations (3.4)–(3.7) are analytically solved by using a MATLAB software.

$$[M^-] = [M^-]_0 \cdot e^{-(k_1+k_2)It} \tag{3.8}$$

$$[M^{-*}] = [M^-]_0 \cdot \frac{k_2}{k_1+k_2-k_3}\left[e^{-k_3 It} - e^{-(k_1+k_2)It}\right] \tag{3.9}$$

$$[e^-(1)] = [M^-]_0 \cdot \frac{k_1}{k_1+k_2}\left[1 - e^{-(k_1+k_2)It}\right] \tag{3.10}$$

$$\left[e^{-}(2)\right] = \left[M^{-}\right]_0 \cdot \left[\frac{k_2}{k_1+k_2} - \frac{k_2}{k_1+k_2-k_3}e^{-k_3It} + \frac{k_2k_3}{(k_1+k_2)(k_1+k_2-k_3)}e^{-(k_1+k_2)It}\right] \quad (3.11)$$

Given that the first-order approximation $e^x \sim 1 + x$ holds when $x \ll 1$, Eqs. (3.8)–(3.11) can be simplified as follows:

$$\left[M^{-}\right] \sim \left[M^{-}\right]_0(1 - (k_1 + k_2)It) \quad (3.12)$$

$$\left[M^{-*}\right] \sim \left[M^{-}\right]_0 \cdot \frac{k_2}{k_1+k_2-k_3}[(1 - k_3It) - (1 - (k_1 + k_2)It)] = k_2It\left[M^{-}\right]_0 \quad (3.13)$$

$$\left[e^{-}(1)\right] \sim \left[M^{-}\right]_0 \cdot \frac{k_1}{k_1+k_2}[1 - (1 - (k_1 + k_2)It)] = k_1It\left[M^{-}\right]_0 \quad (3.14)$$

$$\begin{aligned}\left[e^{-}(2)\right] &\sim \left[M^{-}\right]_0 \cdot \left[\frac{k_2}{k_1 + k_2} - \frac{k_2(1 - k_3It)}{k_1 + k_2 - k_3} + \frac{k_2k_3(1 - (k_1 + k_2)It)}{(k_1 + k_2)(k_1 + k_2 - k_3)}\right] \\ &= \left[M^{-}\right]_0 \cdot \frac{k_2(k_1 + k_2 - k_3) - k_2(k_1 + k_2)(1 - k_3It) + k_2k_3(1 - (k_1 + k_2)It)}{(k_1 + k_2)(k_1 + k_2 - k_3)} = 0\end{aligned} \quad (3.15)$$

when the following approximations are valid:

$$(k_1 + k_2)It \ll 1, k_3It \ll 1. \quad (3.16)$$

Namely, the one-photon photoelectron counts $\left[e^{-}(1)\right]$ at a fixed t is proportional to the laser fluence, whereas the two-photon photoelectron counts $\left[e^{-}(2)\right]$ are negligible in the first-order approximation. Given that the second-order approximation $e^x \sim 1 + x + x^2/2$, Eq. (3.11) can be simplified as follows:

$$\begin{aligned}\left[e^{-}(2)\right] &\sim \left[M^{-}\right]_0 \cdot \left[-\frac{k_2(k_3It)^2/2}{k_1 + k_2 - k_3} + \frac{k_2k_3((k_1 + k_2)It)^2/2}{(k_1 + k_2)(k_1 + k_2 - k_3)}\right] \\ &= \left[M^{-}\right]_0 \cdot \left[\frac{-(k_1 + k_2)k_2k_3^2I^2t^2 + k_2k_3(k_1 + k_2)(k_1 + k_2)I^2t^2}{2(k_1 + k_2)(k_1 + k_2 - k_3)}\right] \\ &= \left[M^{-}\right]_0 \cdot \left[\frac{(k_1+k_2)(k_1+k_2-k_3)k_2k_3I^2t^2}{2(k_1+k_2)(k_1+k_2-k_3)}\right] = \frac{1}{2}k_2k_3I^2t^2\end{aligned}$$

Thus, the two-photon photoelectron counts $\left[e^{-}(2)\right]$ at a fixed t increases quadratically with the laser fluence. On the other hand, when the approximation shown in Eq. (3.16) does not hold, the apparent power numbers of the photoelectron counts $\left[e^{-}(1)\right]$ and $\left[e^{-}(2)\right]$ with respect to the laser fluence are smaller than 1 and 2, respectively. In such a situation, Eq. (3.12) also does not hold and the decrease in the precursor ions $\left[M^{-}\right]_0 - \left[M^{-}\right] = -(k_1 + k_2)It$ is nonlinear. Figure 3.2 shows the double logarithmic plot of the amount of electrons and anions when all reaction rate constants were equal to 1 and the amount of the initial monoanion was equal to 1. The plot qualitatively showed that the slope of the amount of the electron (red

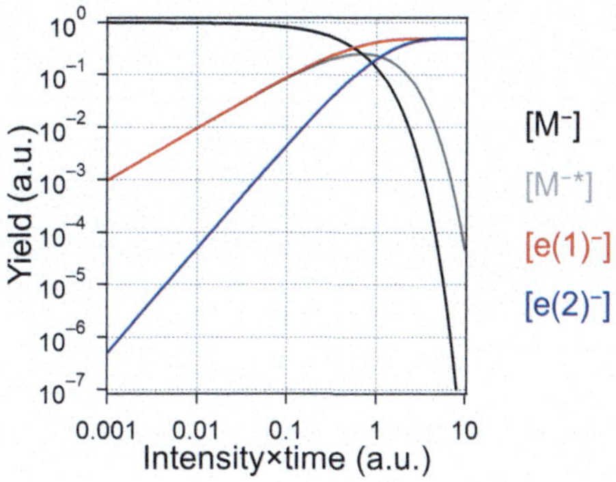

Fig. 3.2 The double logarithmic plot of the amount of anions (black), excited anions (gray), photoelectrons via the one-photon process (red), and those via the two-photon process (blue) simulated based on Eqs. (3.8)–(3.11). Reaction rate constants k_1, k_2, and k_3 and the initial amount of monoanions $[M^-]_0$ were set to 1

and blue trace in Fig. 3.2) is an integer at the low laser fluence while it gradually decreases and approaches zero at the high laser fluence.

3.3 Results and Discussion

3.3.1 Photoelectron Spectroscopy of Gold Superatom

Photoelectron spectra

The PE spectrum of $\mathbf{Au_{13}}^-$ measured at 193 nm (6.42 eV) is shown in Fig. 3.3a. The PE spectrum showed a distinct band at 3.8 eV (referred to as band I) and an intense broad band above 4.4 eV (referred to as band II). These features in the PE spectrum were qualitatively reproduced in the simulated PE spectra obtained based on the result of DFT calculation of $\mathbf{Au_{13}}^-$ with full ligands (Fig. 3.3b). Therefore, Band I and II were assigned to the superatomic 1P orbitals and d band of Au atoms (Fig. 3.3c).

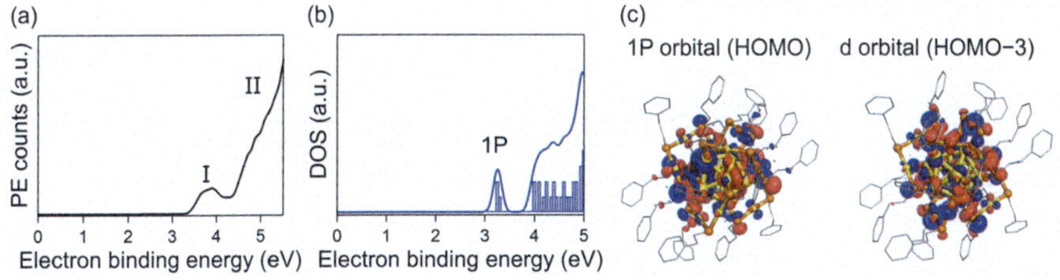

Fig. 3.3 **a** The PE spectrum of $\mathbf{Au_{13}}^-$ measured at 193 nm. **b** The histogram of the energy levels with the bin width of 50 meV (bar) and the simulated PE spectrum (solid line) of $\mathbf{Au_{13}}^-$ obtained by DFT calculation. Each energy level was broadened by using Gauss function with FWHM of 0.2 eV. **c** Kohn–Sham orbitals correspond to superatomic 1P orbitals at HOMO and d orbitals of Au atoms at HOMO−3

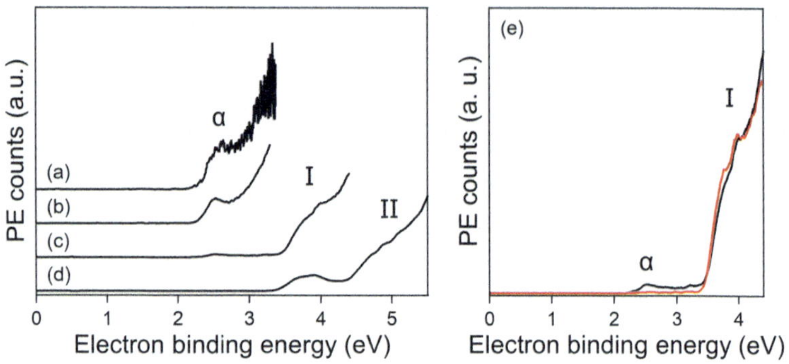

Fig. 3.4 **a** The PE spectrum of $\mathbf{Au_{13}}^-$ reported in the previous report. **b–d** The PE spectra of $\mathbf{Au_{13}}^-$ were recorded at 355, 266, and 193 nm for b–d, respectively. **e** The PE spectra of $\mathbf{Au_{13}}^-$ recorded at 266 nm with the laser fluence of 25 mJ cm^{-2} (black) and 0.5 mJ cm^{-2} (red)

PE spectra of $\mathbf{Au_{13}}^-$ measured at various photon energies are shown in Fig. 3.4. The PE spectrum recorded at 355 nm (3.49 eV) (Fig. 3.4b) exhibited the peak at 2.5 eV (referred to as band α) and reproduced the previously reported PE spectrum (Fig. 3.4a), although photons at 355 nm does not have an enough energy to detach photoelectrons based on Fig. 3.3. The PE spectrum at 266 nm (4.66 eV) showed band α with a tailed structure and the onset region of band I (Fig. 3.4c). The onset energies of bands I and α ($E_{\rm I}$ and E_α) were 3.45 ± 0.06 and 2.32 ± 0.06 eV, respectively. While the E_α value reproduced that reported previously (2.36 eV), the profile of band I differed from that reported previously [14] due to the remarkable improvement of the S/N ratio in the EBE region of >4 eV. The PE spectrum measured at 266 nm in Fig. 3.4c was measured at a fluence of 25 mJ cm^{-2}: the fluence was comparable to that used in the previous report [14]. It was found that the spectral profile recorded at 266 nm was significantly dependent on the laser fluence: the relative intensity of band α with respect to band I decreased at a fluence of 0.5 mJ cm^{-2} (red trace in Fig. 3.4e). This laser fluence dependence suggests that multiple photons were involved in band α.

Investigation of multiphoton process

To examine the multiphoton process in band α, the quantitative assessment of the PE counts against the laser fluence was conducted by attenuating the laser fluence by using a ND filter. The slope of a double logarithmic plot corresponds to the number of photons involved in the PE detachment under the condition that the fraction of the parent ions depleted by photoirradiation is negligibly small (see Sect. 3.2.5 about the principle of the measurement). Experimental results are plotted in Fig. 3.5 and the slopes are summarized in Table 3.2. The reference experiment on I^-, where photoelectron detachment process is known to be one-photon process, showed the linear correlation between the laser fluence and the PE counts in the double logarithmic plot (Fig. 3.5a, b) and the slope was close to 1 (Table 3.2). This result guaranteed the reliability of the measurement. On the other hand, the slope for band α of $\mathbf{Au_{13}}^-$ at

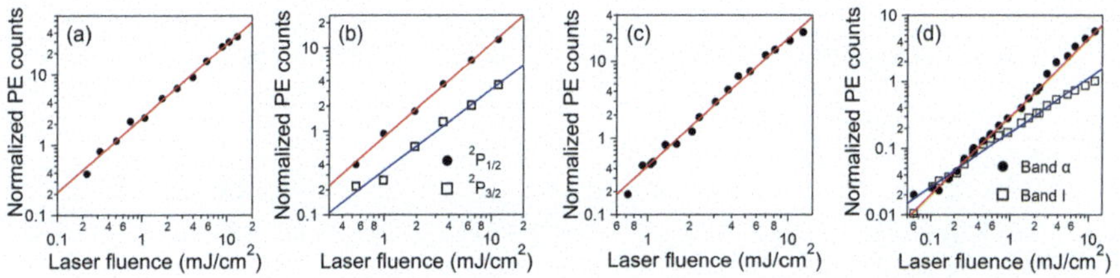

Fig. 3.5 Double logarithmic plots and fitting lines for the PE counts of **a** $^2P_{3/2}$ of I^- at 355 nm, **b** $^2P_{3/2}$ (filled circle) and $^2P_{1/2}$ (hollow square) of I^- at 266 nm, **c** band α of $\mathbf{Au_{13}}^-$ at 355 nm, and **d** band α (filled circle) and band I (hollow square) of $\mathbf{Au_{13}}^-$ at 266 nm plotted against the laser fluences

355 nm was 1.66 ± 0.04 (Fig. 3.5c). The slope was decreased to 1.17 ± 0.02 at 266 nm and considerably smaller than 2, but obviously larger than that for band I (0.82 ± 0.03) (Fig. 3.5d). These results led us to conclude that band α originated from a two-photon process, whereas band I was due to a direct PE detachment (Scheme 3.1a). Non-integer slopes in the double logarithmic plot can be rationalized by the saturation of the photodetachment and photoabsorption process as discussed in Sect. 3.2.5: slopes approach asymptotically from integer values to zero with increasing the laser fluence. In conclusion, we revised the AEA of $\mathbf{Au_{13}}^0$ from 2.36 ± 0.01 to 3.45 ± 0.06 eV (Table 3.1), which is closer to the calculated values (2.94[27] and 3.23 eV[28]) than that reported previously [14].

The most probable scenario for a two-photon process for band α at 355 nm is the PE detachment from a photoexcited state (Scheme 3.1b) as reported in $[Ag_{29}(BDT)_{12}]^{3-}$ (BDT = 1,3-$S_2C_6H_4$) [17]. First, $\mathbf{Au_{13}}^-$ in the S_0 ground state is photoexcited to the S_n state, which is located at the threshold region of the PE detachment. This photoabsorption is followed by a rapid internal conversion (IC) and intersystem crossing (ISC) to $\mathbf{Au_{13}}^{-*}$ in the lowest electronically excited state (T_1). Then, the second photon in the same laser pulse detaches the electron from $\mathbf{Au_{13}}^{-*}$, yielding $\mathbf{Au_{13}}^0$ in the doublet state (D_0). The involvement of the T_1 state in the PE process is supported by the previous reports on the photophysical properties of $\mathbf{Au_{13}}^-$ in solution: the S_n states populated underwent IC and ISC within 10 ps [31, 32] to T_1 with a lifetime of ~10^2 ns [33, 34]. To confirm that band α originated from the electronically excited state, we conducted pump–probe PES on $\mathbf{Au_{13}}^-$ using two nanosecond lasers (Fig. 3.6): $\mathbf{Au_{13}}^-$ was photoexcited to $\mathbf{Au_{13}}^{-*}$ by the pump laser (532 nm), followed by the PE detachment by the probe laser (266 nm). The delay time between pump and probe lights was controlled by the time delay of the trigger pulses

Table 3.2 Slopes of double logarithmic plots in the laser-fluence measurement

Band	I^-		Au_{13}^-	
	$^2P_{1/2}$	$^2P_{3/2}$	Band α	Band I
355 nm	1.06 ± 0.03	–	1.66 ± 0.04	–
266 nm	1.11 ± 0.03	0.97 ± 0.06	1.17 ± 0.02	0.82 ± 0.03

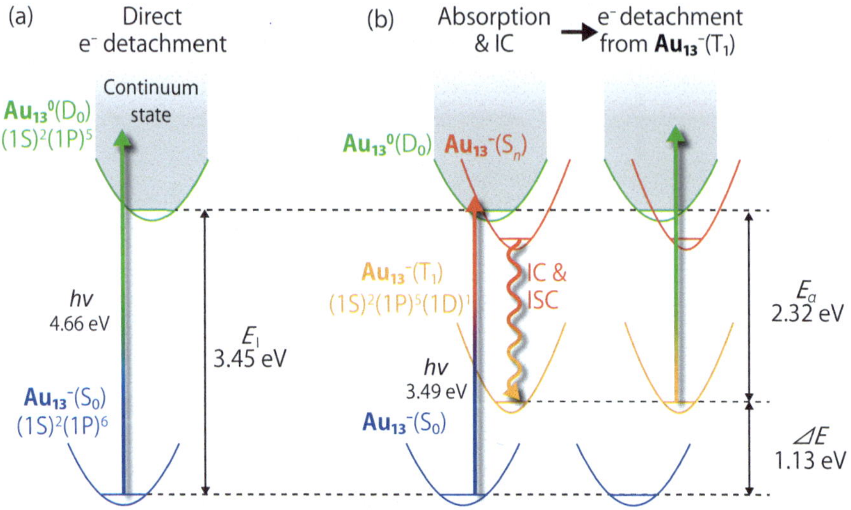

Scheme 3.1 Schematic illustration of the PE detachment of $\mathbf{Au_{13}}^-$ by **a** one-photon process at 266 nm and **b** two-photon process at 355 nm. For simplicity, panel (**b**) depicts only the IC and ISC process between the vibrationally ground states and the subsequent PE detachment processes

to the pump and probe lasers. The intensity of band α increased when the pump and probe lights overlapped in time and decayed with the delay time of the probe light with respect to the pump light (Fig. 3.6a). This result indicates that $\mathbf{Au_{13}}^{-*}$ populated by the pump laser had a lifetime longer than a nanosecond. In contrast, for naked Au [35, 36] and Ag [37] cluster anions, the PE detachment via electronically excited states was rarely observed using a nanosecond laser but was observed with a femtosecond laser. The elongated lifetimes of the electronically excited states of $\mathbf{Au_{13}}^-$ may have originated from the suppression of nonradiative relaxation by ligation-induced core rigidification, as supported by the emergence of the photoluminescence [38, 39].

The interpretation of the two-photon PE spectra, including band α and weak signals between bands α and I, is not simple because important factors such as the shape of the potential energy surface and the population of vibrational states are unknown. Hot band profile was not observed in the lower EBE region of band α although $\mathbf{Au_{13}}^-(T_1)$ is more likely to be populated in the vibrationally excited states due to lower efficiency of vibrational cooling in the gas phase than in solution. This suggests that band α is assigned to the PE detachment from the vibrationally ground state of the T_1 state. Under this assumption, the energy difference (ΔE) between E_α and E_I corresponds to that between the vibrationally ground states of $\mathbf{Au_{13}}^-(T_1)$ and $\mathbf{Au_{13}}^-(S_0)$ (Scheme 3.1b). The ΔE value of 1.13 eV (Table 3.3) was smaller than the optical HOMO–LUMO gap (E_HL) of 1.36 eV determined from the onset of the absorption spectrum (Fig. 3.1b), which is consistent with the fact that the T_1 state of $\mathbf{Au_{13}}^-$ is more stable than the S_1 state according to the phosphorescence measurement [32]. Within the framework of this model, the weak PE signals between

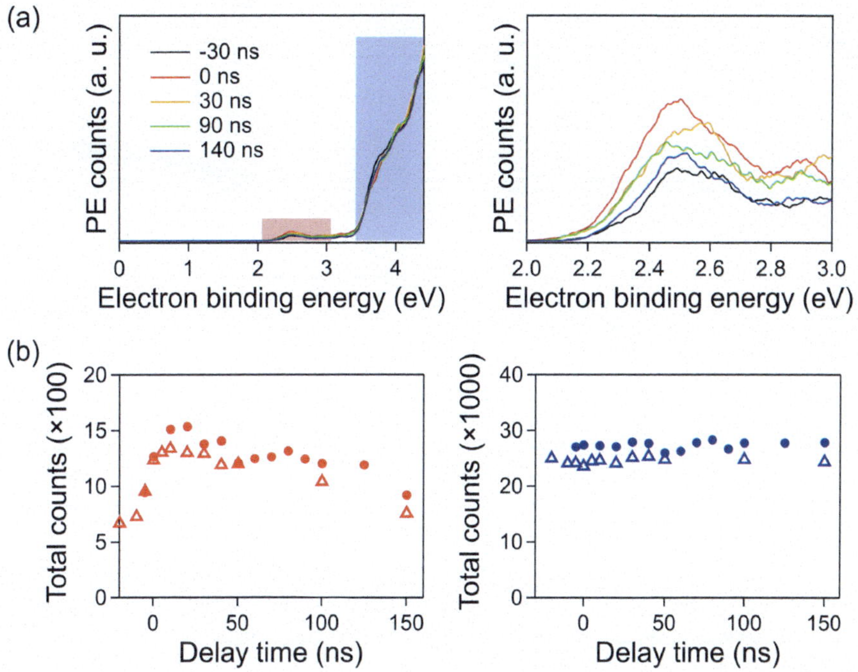

Fig. 3.6 **a** PE spectra of $\mathbf{Au_{13}}^-$ measured at 266 nm (5 mJ cm^{-2}) while irradiating 532 nm (30 mJ cm^{-2}) from another Nd:YAG laser at different timings. Right spectra show the enlarged views. **b** PE counts plotted against the delay time of two laser pulses. Left and right panels show the PE counts of band A (red region in panel a) and band B (blue reigon in panel a). Filled circles and hollow triangles represent the results of different experiments. The delay time in the figure was set to zero when the PE counts of band A were increased. This timing was close to the simultaneous irradiation of two laser pulses roughly monitored by a photodiode sensor

Table 3.3 Summary of energetics

Cluster	E_α (eV)	E_I (eV)	ΔE (eV)[a]	E_HL (eV)[b]
$[Au_{25}(PET)_{18}]^-$ ($\mathbf{Au_{13}}^-$)	2.32	3.45	1.13	1.36
$[Ag_{25}(DMBT)_{18}]^-$ ($\mathbf{Ag_{13}}^-$)	2.66	3.97	1.31	1.54
$[AuAg_{24}(DMBT)_{18}]^-$ ($\mathbf{AuAg_{12}}^-$)	2.52	4.00	1.48	1.60
$[PdAg_{24}(DMBT)_{18}]^{2-}$ ($\mathbf{PdAg_{12}}^{2-}$)	0.67	2.00	1.33	1.59
$[PtAg_{24}(DMBT)_{18}]^{2-}$ ($\mathbf{PtAg_{12}}^{2-}$)	0.56	2.08	1.52	1.68

[a] $\Delta E = E_\mathrm{I} - E_\alpha$. [b] Determined by the onset of the optical spectra (Fig. 3.1b)

bands α and I are assignable to PE detachment from the vibrationally excited states of $\mathbf{Au_{13}}^-(S_0)$ in the electronically ground state populated by the IC from $\mathbf{Au_{13}}^-(T_1)$.

Another possible two-photon process is the PE detachment from fragmented anions generated by the first photon [40]. To check such a possibility, the photodissociation (PD) mass spectrum of $\mathbf{Au_{13}}^-$ was recorded by irradiating a 266 nm laser 1

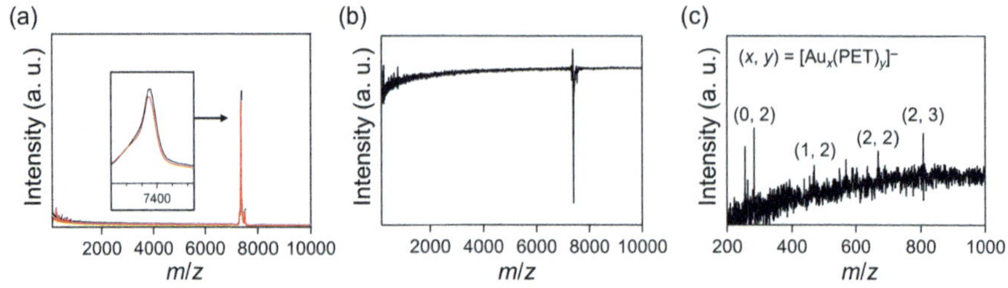

Fig. 3.7 **a** ESI mass spectra of $\mathbf{Au_{13}}^-$ without (black) and with (red) photodissociation laser at 266 nm. **b** Difference spectra of (**a**). Negative and positive peaks indicate the depletion and the generation of ions, respectively. **c** Enlarged spectra of (**b**)

μs before the pulsed acceleration of ions at the TOF–MS (Fig. 3.6a). The differential mass spectra with and without the PD laser (Fig. 3.7) showed the depletion of $\mathbf{Au_{13}}^-$ and the generation of a small amount (<10% of the $\mathbf{Au_{13}}^-$ depletion) of $[\mathrm{Au}_x(\mathrm{PET})_y]^-$ with $(x, y) = (0, 2)$, (1, 2), (2, 2), and (2, 3). This result suggests that the PD was a minor process compared to the PE detachment. In addition, the E_α value (2.32 eV) could not be explained by the AEAs of $[\mathrm{Au(SCH_3)_2}]^0$ and $[\mathrm{Au_2(SCH_3)_3}]^0$ (3.33 and 4.06 eV, respectively) [41]. Thus, we concluded that band α was not assigned to the PE signals from photofragments, but to those from $\mathbf{Au_{13}}^{-*}$ via the following two-step process:

$$\mathbf{Au}_{13}^{-} + h\nu \rightarrow \mathbf{Au}_{13}^{-*} \tag{3.18}$$

$$\mathbf{Au}_{13}^{-*} + h\nu \rightarrow \mathbf{Au}_{13}^{0} + e^{-}. \tag{3.19}$$

PE spectra of other ligand-protected metal clusters may also involve the PE detachment via photoexcited states since the lifetimes are generally longer than the nanosecond order. Therefore, the dependence of the PE spectrum on the laser wavelength and fluence should at least be studied in order to assign a direct PE signal.

Two-photon PE detachment of ligand-protected metal clusters via photoexcited states is applicable to investigate the excited state dynamics and suggests the potential application in nonlinear optical processes. The long excited-state lifetime of ligand-protected metal clusters may also be suitable to conduct resonant-enhanced multiphoton ionization spectroscopy to evaluate the deep energy levels such as superatomic 1S orbital far below 1P orbitals.

3.3.2 Photoelectron Spectroscopy of Silver-Based Superatoms

The PE spectra of the other singly charged anions $\mathbf{Ag_{13}}^-$ and $\mathbf{AuAg_{12}}^-$ are shown in Fig. 3.8a, b, respectively. At 266 nm, the spectra of both anions exhibited a main band in the EBE region of >4 eV (band I) with a small band at the lower EBE region (band α) at a high laser fluence of 25 mJ cm^{-2}. The E_α and E_I values are summarized in Table 3.3. As in the case of $\mathbf{Au_{13}}^-$, the relative intensity of band α to band I was smaller at a lower laser fluence (insets of Fig. 3.8). When $\mathbf{Ag_{13}}^-$ was irradiated at 355 nm with a fluence of 150 mJ cm^{-2}, the E_α value was further reduced to 2.07 eV, which agrees with the reported AEA value [15]. These results indicate that band α observed in the low EBE region reported here and previously [15] is associated with multiphoton processes as described in Eqs. (3.18) and (3.19). A comparison with the PE spectrum at 193 nm confirmed that band I of $\mathbf{Ag_{13}}^-$ in Fig. 3.8a was due to the PE signals from the HOMO (1P superatomic orbital): the AEAs of $\mathbf{Ag_{13}}^0$ and $\mathbf{AuAg_{12}}^0$ were redetermined as 3.97 and 4.00 eV, respectively, from the corresponding E_I values (Table 3.1). The newly determined AEAs of $\mathbf{Ag_{13}}^0$ and $\mathbf{AuAg_{12}}^0$ are much larger than those previously reported (2.02 and 2.08 eV[15]) and are closer to the calculated values (3.46 and 3.51 eV[15]). The ΔE values of $\mathbf{Ag_{13}}^-$ and $\mathbf{AuAg_{12}}^-$ are slightly smaller than the corresponding E_HL values (Table 3.3) as well as that of $\mathbf{Au_{13}}^-$.

The PE spectra of doubly charged anions $\mathbf{PdAg_{12}}^{2-}$ and $\mathbf{PtAg_{12}}^{2-}$ recorded at 266 nm are presented in Fig. 3.9. In both cases, the PE signals were suppressed in the high EBE region because the PE detachment from deeper orbitals was hindered by the RCB created by electrostatic repulsion between the detached electron and the remaining anion as introduced in Sect. 1.3 [25, 26]. The heights of the RCB are estimated to be ~1.4 eV from the difference between the photon energy (4.66 eV) and the EBE value at the top of band II. The PE spectra of $\mathbf{PdAg_{12}}^{2-}$ and $\mathbf{PtAg_{12}}^{2-}$ are similar in shape, indicating the similar electronic structures between

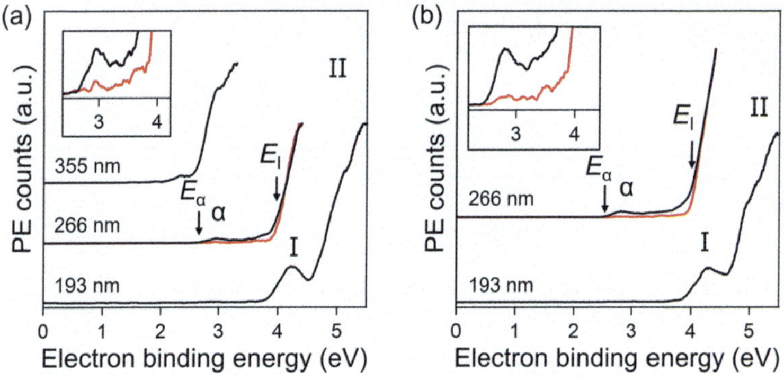

Fig. 3.8 PE spectra of **a** $\mathbf{Ag_{13}}^-$ and **b** $\mathbf{AuAg_{12}}^-$. Black and red traces at 266 nm were recorded at a laser fluence of 25 and 0.5 mJ cm^{-2}, respectively

the Pd@Ag_{12} and Pt@Ag_{12} superatoms. The PE spectra of $\mathbf{PdAg_{12}}^{2-}$ and $\mathbf{PtAg_{12}}^{2-}$ exhibit a main band with a peak at an EBE of ~3.3 eV (band II) and a hump (band I) with the E_I values of 2.00 and 2.08 eV, respectively (Table 3.3). Bands I and II are assigned to the PE detachment from the 1P superatomic orbital and Ag(4d) bands, respectively. On the other hand, band α with the onset energies of 0.67 and 0.56 eV at 355 nm and band β with the onset energy of ~4 eV at 266 nm were observed in the spectra recorded at a fluence of 25 mJ cm^{-2} (black traces in Fig. 3.9), as reported previously [15]. The laser fluence dependence indicates that bands α and β in the spectra of $\mathbf{PdAg_{12}}^{2-}$ and $\mathbf{PtAg_{12}}^{2-}$ are associated with two-photon processes. Band α is assigned to PE detachment from photoexcited states (Scheme 3.2a), as in the case of $\mathbf{Au_{13}}^{-}$, $\mathbf{Ag_{13}}^{-}$, and $\mathbf{AuAg_{12}}^{-}$ (Scheme 3.1b and Eqs. (3.18) and (3.19)). Band β cannot be assigned to the PE detachment from doubly charged anions because the RCB with a height of ~1.4 eV prohibits the emission of such slow electrons. The most probable assignment of band β is PE detachment from singly charged anions such as $\mathbf{PdAg_{12}}^{-}$ and $\mathbf{PtAg_{12}}^{-}$ produced by the PE detachment of $\mathbf{PdAg_{12}}^{2-}$ and $\mathbf{PtAg_{12}}^{2-}$, respectively (Scheme 3.2b):

$$\mathbf{PdAg_{12}^{2-}} + h\nu \rightarrow \mathbf{PdAg_{12}^{-}} + e^{-} \tag{3.20}$$

$$\mathbf{PdAg_{12}^{-}} + h\nu \rightarrow \mathbf{PdAg_{12}^{0}} + e^{-}. \tag{3.21}$$

In conclusion, we redetermined the AEAs of $\mathbf{PdAg_{12}}^{-}$ and $\mathbf{PtAg_{12}}^{-}$ to be 2.00 and 2.08 eV, respectively (Table 3.1), from the E_I values. The newly determined AEAs of $\mathbf{PdAg_{12}}^{-}$ and $\mathbf{PtAg_{12}}^{-}$ support the theoretical prediction that the AEA of $\mathbf{Ag_{13}}^{0}$ (3.46 eV) is reduced by ~ 2.2 eV upon the doping of a single Pd or Pt atom.

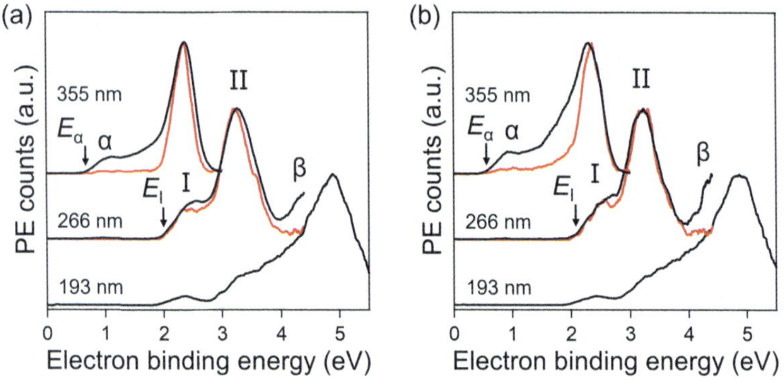

Fig. 3.9 PE spectra of **a** $\mathbf{PdAg_{12}}^{2-}$ and **b** $\mathbf{PtAg_{12}}^{2-}$. Black and red traces at 355 and 266 nm were recorded at a laser fluence of 25 and 0.5 mJ cm^{-2}, respectively

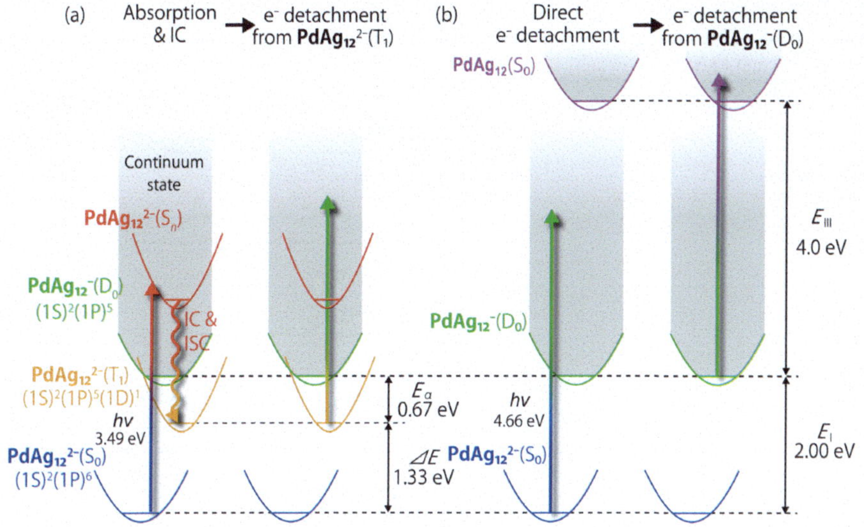

Scheme 3.2 Schematic illustration of the two-photon PE detachment of $\mathbf{PdAg_{12}}^{2-}$ via $\mathbf{PdAg_{12}}^{2-*}$ in an electronically excited state and **b** singly charged anion $\mathbf{PdAg_{12}}^{-}$. For simplicity, panel (**a**) depicts only the IC and ISC process between the vibrationally ground states and the subsequent PE detachment processes

3.3.3 Effect of Heterometal Doping

PE spectra of $\mathbf{Au_{13}}^{-}$, $\mathbf{Ag_{13}}^{-}$, $\mathbf{AuAg_{12}}^{-}$, $\mathbf{PdAg_{12}}^{2-}$, and $\mathbf{PtAg_{12}}^{2-}$ measured at 193 nm were listed in Fig. 3.10. All PE spectra exhibited one single distinct peak at the spectral onset assigned to the superatomic 1P orbitals. The spectral profiles of $\mathbf{Ag_{13}}^{-}$ and $\mathbf{AuAg_{12}}^{-}$ are quite similar, indicating that the electronic structure of the Ag_{13} superatom is not appreciably affected by the replacement of Ag at center with Au. The electronic structure looks qualitatively similar even after the complete metal exchange as shown in the PE spectra of $\mathbf{Au_{13}}^{-}$ and $\mathbf{Ag_{13}}^{-}$, while the spectral shift to the high binding energy was observed with the silver superatom. It is noteworthy that corresponding neutral superatoms with 7 electrons $\mathbf{Au_{13}}^{0}$, $\mathbf{Ag_{13}}^{0}$, and $\mathbf{AuAg_{12}}^{0}$ can be categorized as superhalogens because their AEA values are comparable or larger than those of halogen atoms (3.06–3.62 eV) [42] and a well-known superhalogen Al_{13} (3.57 eV) [43–45]. The halogen-like electron affinities suggest that these superatoms can be served as alternative halogens with the structural tunability for functionalization.

PE spectra of $\mathbf{PdAg_{12}}^{2-}$, and $\mathbf{PtAg_{12}}^{2-}$ were largely different from those of singly charged anions $\mathbf{Ag_{13}}^{-}$. A remarkable red-shift of the 1P orbitals by ~2 eV with respect to those of $\mathbf{Ag_{13}}^{-}$ can be ascribed to the increase of the total negative charge from −1 to −2. In a simple jellium model, the additional negative charge induced the upshift of the jellium potential (Fig. 3.11). In addition, the different valency between Pd and Ag resulted in the non-uniform potential according to the two-step spherical jellium

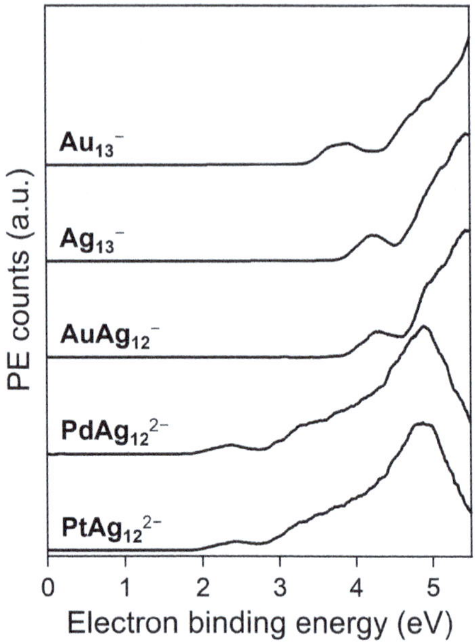

Fig. 3.10 PE spectra of $\mathbf{Au_{13}}^-$, $\mathbf{Ag_{13}}^-$, $\mathbf{AuAg_{12}}^-$, $\mathbf{PdAg_{12}}^{2-}$, and $\mathbf{PtAg_{12}}^{2-}$ measured at 193 nm

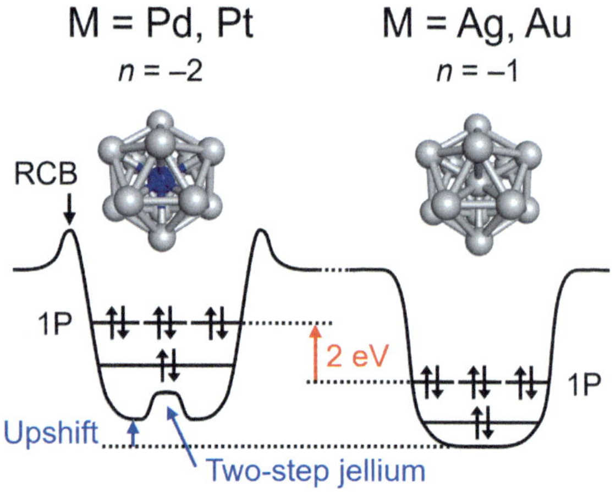

Fig. 3.11 Schematic illustration of the jellium potential of $\mathbf{MAg_{12}}^{2-}$ (M = Pd, Pt) and $\mathbf{Ag_{13}}^-$ (M = Ag, Au)

model [46, 47]. In the two-step jellium model, the less electronegative heteroatom at the center of the cluster makes the bump at the center of the effective potential. In the case of $\mathbf{Ag_{13}}^-$ and $\mathbf{PdAg_{12}}^{2-}$, the charge state of the central atom before filling the electron is formally Ag^+ and Pd^0 and thus the center of the effective potential of $PdAg_{12}$ core is upshifted than Ag_{13} core. For the same group elements having the same valency, the Pauling electronegativity (EN) of elements can qualitatively describe the shape of the effective potential. The central doping of Au with the EN of 2.54 into Ag with the EN of 1.93 induces the attractive potential at the center of the cluster and leads to the larger AEA of $\mathbf{Au_{13}}^0$ (4.00 eV) than $\mathbf{Ag_{13}}^0$ (3.97 eV). The doping effect was qualitatively similar to the previous report based on the PE spectra with two-photon PE detachment signals via photoexcited states due to the small doping effect on E_{HL} [15]. This study shows the intrinsic effect of constituent elements and charges on the electronic structures of the superatoms.

3.4 Summary

In summary, we revisited the AEAs of structurally defined superatoms with eight electrons by using an improved PES apparatus with highly intensified ion beam. It was found that the AEAs reported previously were underestimated due to the involvement of PE detachment via photoexcited states. Thus, the newly determined AEAs were 1–2 eV larger than those reported previously, and the qualitative effects of the doping on the AEAs of the superatoms were not affected. This study provides intrinsic information on the electronic structures and photoinduced processes of ligand-protected metal superatoms, which is useful for explaining and designing their specific properties.

References

1. Tsukuda T, Häkkinen H (2015) Protected metal clusters: from fundamentals to applications, 1st edn. Amsterdam, The Netherlands, Elsevier B.V.
2. Jin R, Zeng C, Zhou M, Chen Y (2016) Chem Rev 116:10346
3. Chakraborty I, Pradeep T (2017) Chem Rev 117:8208
4. Yao Q, Yuan X, Chen T, Leong DT, Xie J (2018) Adv Mater 30:1802751
5. Du Y, Sheng H, Astruc D, Zhu M (2020) Chem Rev 120:526
6. Kang X, Li Y, Jin M, Zhu R (2020) Chem Soc Rev 49:6443
7. Omoda T, Takano S, Tsukuda T (2020) Small 16:2001439
8. Hirai H, Ito S, Takano S, Koyasu K, Tsukuda T (2020) Chem Sci 11:12233
9. Kawawaki T, Ebina A, Hosokawa Y, Ozaki S, Suzuki D, Hossain S, Negishi Y (2021) Small 17:2005328
10. Takano S, Tsukuda T (2021) J Am Chem Soc 143:1683
11. Walter M, Akola J, Acevedo L-O, Jadzinsky PD, Calero G, Ackerson CJ, Whetten RL, Grönbeck H, Häkkinen H (2008) Proc Natl Acad Sci 105:9157
12. Jena P, Sun Q (2021) Superatoms: principles, synthesis and applications, 1st edn. Wiley, New York
13. Hirata K, Yamashita K, Muramatsu S, Takano S, Ohshimo K, Azuma T, Nakanishi R, Nagata T, Yamazoe S, Koyasu K, Tsukuda T (2017) Nanoscale 9:13409
14. Hirata K, Kim K, Nakamura K, Kitazawa H, Hayashi S, Koyasu K, Tsukuda T (2019) J Phys Chem C 123:13174
15. Kim K, Hirata K, Nakamura K, Kitazawa H, Hayashi S, Koyasu K, Tsukuda T (2019) Angew Chem Int Ed 58:11637
16. Hirata K, Tomihara R, Kim K, Koyasu K, Tsukuda T (2019) Phys Chem Chem Phys 21:17463
17. Veenstra AP, Monzel L, Baksi A, Czekner J, Lebedkin S, Schneider EK, Pradeep T, Unterreiner A-N, Kappes MM (2020) J Phys Chem Lett 11:2675

18. Tasaka Y, Nakamura K, Malola S, Hirata K, Kim K, Koyasu K, Häkkinen H, Tsukuda T (2020) J Phys Chem Lett 11:3069
19. Koyasu K, Tsukuda T (2021) J Chem Phys 154:140901
20. Heaven MW, Dass A, White PS, Holt KM, Murray RW (2008) J Am Chem Soc 130:3754
21. Zhu M, Aikens CM, Hollander FJ, Schatz GC, Jin R (2008) J Am Chem Soc 130:5883
22. Joshi CP, Bootharaju MS, Alhilaly MJ, Bakr OM (2015) J Am Chem Soc 137:11578
23. Yan J, Su H, Yang H, Malola S, Lin S, Häkkinen H, Zheng N (2015) J Am Chem Soc 137:11880
24. Bootharaju MS, Joshi CP, Parida MR, Mohammed OF, Bakr OM (2016) Angew Chem Int Ed 551922
25. Wang L-S, Wang X-B (2000) J Phys Chem A 104:1978
26. Dau PD, Liu H-T, Yang J-P, Winghart M-O, Wolf TJA, Unterreiner A-N, Weis P, Miao Y-R, Ning C-G, Kappes MM, Wang L-S (2012) Phys Rev A 85:064503
27. Jung J, Kang S, Han Y-K (2012) Nanoscale 4:4206
28. Matus MF, Malola S, Bonilla KE, Barngrover BM, Aikens CM, Häkkinen H (2020) Chem Commun 56:8087
29. Tofanelli MA, Salorinne K, Ni TW, Malola S, Newell B, Phillips B, Häkkinen H, Ackerson CJ (2016) Jahn-Teller effects in $Au_{25}(SR)_{18}$. Chem Sci 7:1882
30. Frisch MJ, Trucks GW, Schlegel HB, Scuseria GE, Robb MA, Cheeseman JR, Scalmani G, Barone V, Petersson GA, Nakatsuji H, Li X, Caricato M, Marenich AV, Bloino J, Janesko BG, Gomperts R, Mennucci B, Hratchian HP, Ortiz JV, Izmaylov AF, Sonnenberg JL, Williams-Young D, Ding F, Lipparini F, Egidi F, Goings J, Peng B, Petrone A, Henderson T, Ranasinghe D, Zakrzewski VG, Gao J, Rega N, Zheng G, Liang W, Hada M, Ehara M, Toyota K, Fukuda R, Hasegawa J, Ishida M, Nakajima T, Honda Y, Kitao O, Nakai H, Vreven T, Throssell K, Montgomery Jr. JA, Peralta JE, Ogliaro F, Bearpark MJ, Heyd JJ, Brothers EN, Kudin KN, Staroverov VN, Keith TA, Kobayashi R, Normand J, Raghavachari K, Rendell AP, Burant JC, Iyengar SS, Tomasi JJ, Cossi M, Millam JM, Klene M, Adamo C, Cammi R, Ochterski JW, Martin RL, Morokuma K, Farkas O, Foresman JB, Fox DJ (2016) Gaussian 16, Revision B.01. Gaussian, Inc., Wallingford CT
31. Green TD, Knappenberger KL (2012) Nanoscale 4:4111
32. Liu Z, Zhou M, Luo L, Wang Y, Kahng E, Jin R (2023) J Am Chem Soc 145:19969
33. Wu Z, Jin R (2010) Nano Lett 10:2568
34. Zhou M, Song Y (2021) J Phys Chem Lett 12:1514
35. Niemietz M, Gerhardt P, Ganteför G, Kim YD (2003) Chem Phys Lett 380:99
36. Walter M, Häkkinen H, Stanzel J, Neeb M, Eberhardt W (2007) Phys Rev B 76:155422
37. Niemietz M, Engelke M, Kim YD, Ganteför G (2007) Phys Rev B 75:085438
38. Pyo K, Thanthirige VD, Kwak K, Pandurangan P, Ramakrishna G, Lee D (2015) J Am Chem Soc 137:8244
39. Kang X, Zhu M (2019) Chem Soc Rev 48:2422
40. Ganteför G, Kraus S, Eberhardt W (1998) J Electron Spectrosc Relat Phenom 88–91:35
41. Ning C-G, Xiong X-G, Wang Y-L, Li J, Wang L-S (2012) Phys Chem Chem Phys 14:9323
42. Gutsev GL, Boldyrev AI (1981) Chem Phys 56:277
43. Li X, Wang L-S (2002) Phys Rev B 65:153404
44. Bergeron DE, Castleman AW, Morisato T, Khanna SN (2004) Science 304:84
45. Reber AC, Khanna SN, Castleman AW (2007) J Am Chem Soc 129:10189
46. Baladron C, Alonso JA (1988) Phys B 154:73
47. Janssens E, Neukermans S, Lievens P (2004) Curr Opin Solid State Mater Sci 8:185

Chapter 4
Effect of Ligand Substitution on Electron Affinities of Eight Electron Superatoms

Abstract The fourth chapter describes the effect of stepwise ligand substitution and the role of the electron-withdrawing alkynyl ligands on the eight-electron superatom. The role of alkynyl ligands with electron-withdrawing nature in the stability of metal clusters was investigated by gas-phase anion photoelectron spectroscopy on structurally defined superatoms modified with the stepwise ligand substitution. Anion photoelectron spectroscopy on $[PdAu_{24}(C{\equiv}CAr^F)_{18-x}(C{\equiv}CPh)_x]^{2-}$ ($Ar^F = 3,5\text{-}(CF_3)_2C_6H_3$) with specific x ($= 0 - 6$) revealed that electron binding energies decreased linearly with x, indicating that the electron-withdrawing CF_3 substituents on the alkynyl ligand played a critical role in the electronic stabilization of $[PdAu_{24}(C{\equiv}CAr^F)_{18}]^{2-}$. Density functional theory calculations reproduced the decrease of electron binding energies and the ligand effect was ascribed to the electrostatic effect of the dipole moment of ligands, which is a similar mechanism to the modulation of the work function of gold films by organic monolayers.

Keywords Electron-withdrawing · Superatom · Gold cluster · Density functional theory calculation

4.1 Introduction

Atomically precise gold clusters protected by organic ligands are attractive nanomaterials with novel physicochemical properties that are non-scalable from those of the bulk gold [1, 2]. Since the first report of alkynyl-protected Au clusters $Au_x(C{\equiv}CR)_y$ in 2011 [3], their stable compositions [3–10], geometric structures [11], optical properties [3, 5], and catalytic activities [12, 13] have been studied extensively. These studies have revealed that $Au_x(C{\equiv}CR)_y$ exhibit several notable features distinct from those of the well-studied thiolate-protected Au clusters $Au_x(SR)_y$ [14] even though they share the same stable compositions (x, y). A richer diversity of interfacial structures is one of the unique features of $Au_x(C{\equiv}CR)_y$ [11]. Further differences are the electronic conjugation between R groups and the Au core in $Au_x(C{\equiv}CR)_y$ [15, 16] and that electron-withdrawing C≡CR ligands such as fluorinated phenylacetylenes

S. Ito, *Electronic Structures and Bonding Interaction of Structurally Defined Gold/Silver Superatoms*, Springer Theses, https://doi.org/10.1007/978-981-96-2892-6_4

are favored for the synthesis [4, 5, 8, 10]. For example, $[Au_{25}(C{\equiv}CR)_{18}]^-$ can be synthesized only by using electron-withdrawing ligands [5, 17] whereas ubiquitous $[Au_{25}(SR)_{18}]^-$ can be synthesized regardless of the R groups [18]. This difference suggests that electron-withdrawing C≡CR ligands play an essential role in the stabilization of $Au_x(C{\equiv}CR)_y$. One of the possible roles may be the electronic stabilization as revealed in $[Au_{25}(SR)_{18}]^-$ by the electrochemical measurements and density functional theory (DFT) calculations [19–22].

In this work, I investigated how the electronic structure of $[PdAu_{24}(C{\equiv}CAr^F)_{18}]^{2-}$ (Ar^F = 3,5-$(CF_3)_2C_6H_3$, Fig. 4.1, referred to as $\mathbf{PdAu_{12}}^{2-}$) [17, 23] evolves by stepwise substitution of $C{\equiv}CAr^F$ with non-fluorinated C≡CPh. I chose $[PdAu_{24}(C{\equiv}CAr^F)_{18}]^{2-}$ rather than $[Au_{25}(C{\equiv}CAr^F)_{18}]^-$ because the large electron binding energy of the latter [6] is reduced to an accessible level due to the Pd doping and increase of the total negative charge. The precursor cluster $\mathbf{PdAu_{12}}^{2-}$ consists of an icosahedral $Pd@Au_{12}$ core protected by six bidentate $Au_2(C{\equiv}CAr^F)_3$ oligomers (Fig. 4.1a). According to the two-step jellium model [24, 25], eight valence electrons are accommodated into 1S and 1P superatomic orbitals [26] created within the deformed potential well provided by the $Pd@Au_{12}$ core and the repulsive Coulomb barrier (RCB) [27] formed by the electrostatic repulsion between $\mathbf{PdAu_{12}}^{-}$ and an electron (Fig. 4.1b). As a result, $\mathbf{PdAu_{12}}^{2-}$ can be regarded as a structurally defined superatom with the closed-shell electron configuration $(1S)^2(1P)^6$. The electronic structures of ligand-substituted superatoms $[PdAu_{24}(C{\equiv}CAr^F)_{18-x}(C{\equiv}CPh)_x]^{2-}$ ($\mathbf{PdAu_{12}/}\boldsymbol{x}^{2-}$) were evaluated by gas-phase anion photoelectron spectroscopy (PES) [28–32]. According to Koopmans' theorem, PES provides the electron binding energies (EBEs) of the 1P superatomic orbitals, corresponding to the highest occupied molecular orbitals (HOMOs), with respect to the vacuum level, without perturbation from solvents and counter ions [33–36]. PE spectra of mass-selected $\mathbf{PdAu_{12}/}\boldsymbol{x}^{2-}$ (x = 0–6) revealed that EBEs of 1P orbitals decreased linearly with respect to x by ~75 meV per one ligand exchange in the range of x = 0–6. Similar electronic stabilization by electron-withdrawing ligands was also observed in the electrochemical measurements of $\mathbf{PdAu_{12}/}\boldsymbol{x}^{2-}$ as a function of the average value of x. Based on DFT calculations I proposed that the CF_3 substituent groups of the alkynyl ligands stabilized the superatomic orbitals via a similar mechanism as the self-assembled monolayers (SAMs) modified the work function of bulk materials.

4.2 Methods

4.2.1 General

All solvents except for tetrahydrofuran, sodium borohydride ($NaBH_4$), and triethylamine were purchased from FUJIFILM Wako Pure Chemical Corporation. 3,5-Bis(trifluoromethyl)ethynylbenzene (3,5-$(CF_3)_2C_6H_3C{\equiv}CH$, $Ar^FC{\equiv}CH$) and

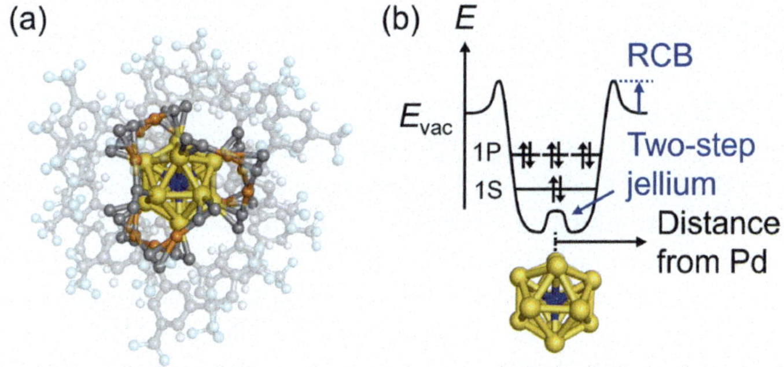

Fig. 4.1 **a** Crystal structure and **b** schematic electronic structure of $\mathbf{PdAu_{12}}^{2-}$. Color codes: yellow = Au in core, orange = Au in oligomers, blue = Pd, gray = C, white = H, cyan = F. Ar^F moieties are shown as transparent for the clarity

tetraphenylphosphonium chloride (PPh_4Cl) were purchased from Tokyo Chemical Industry. Tetrahydrofuran (THF) and phenylacetylene (PhC≡CH) were purchased from Kanto Chemicals. $[PdAu_8(PPh_3)_8](NO_3)_2$ and $(Me_2S)AuCl$ were synthesized according to the reported procedures [17, 37–39]. UV–vis absorption spectra were recorded by a JASCO V-670 spectrometer using methanol as a solvent. Electrospray ionization mass spectrometry (ESI–MS) was performed by using a JEOL JMS-T100LC time-of-flight mass spectrometer. Typical conditions of ESI–MS were as follows: flow rate = 3 mL/h, spray voltage = –2.5 kV, ring lens voltage = –40 V, orifice 1 voltage = –50 V, orifice 2 voltage = 0 V, desolvation chamber temperature = 120 °C, orifice 1 temperature = 80 °C. Obtained mass spectra were calibrated using $[(CsI)_nI]^-$ as an external reference. Voltammetric measurements were conducted using a BAS model 720B potentiostat equipped with a Faraday cage. Recrystallized $TBAPF_6$ was used as a supporting electrolyte. Crystals of the samples were directly dissolved in dichloromethane at a concentration of 0.1 M, and the solution was bubbled with Ar gas for 5 min before the measurements. The electrochemical cell consisted of a Pt wire counter electrode, an Ag wire soaked in the supporting electrolyte solution separated by ion permeable glass as a reference electrode, a glassy carbon working electrode (φ1 mm), an Ar gas inlet, and a gas outlet. The glassy carbon electrode was successively polished using 0.05 μm Al_2O_3 paste. The polished electrode was washed with water and EtOH and sonicated in EtOH for 5 min. After sonication, the electrode was dried under a stream of Ar. Differential pulse voltammograms (DPVs) were recorded under the following conditions: Potential increments: 4 mV; pulse amplitude: 50 mV; pulse width: 50 ms; sampling width: 16.7 ms; pulse period: 100 ms.

4.2.2 *Synthesis*

Synthesis of $[PdAu_{24}(C{\equiv}CAr^F)_{18}](PPh_4)_2$ was conducted based on the previous report [17]. $(Me_2S)AuCl$ (35.8 mg, 120 μmol) was suspended in THF (2 mL) and stirred in the dark. $Ar^FC{\equiv}CH$ (15.3 μL, 90 mol) and triethylamine (12.6 μL, 90 μmol) were added, resulting in a colorless solution (solution **A**) containing Au–C≡CR polymer complexes. $[PdAu_8(PPh_3)_8](NO_3)_2$ (19.4 mg, 5 μmol) was dissolved in a mixed solution of EtOH (500 μL) and THF (5 mL) in another test tube and the solution was cooled in an ice bath. Then, an ethanolic solution of $NaBH_4$ (0.10 M, 50 μL) was added to this chilled solution and after 30 s, solution **A** was added dropwise over 2–3 min. After stirring for 30 min, PPh_4Cl (3.7 mg, 10 μmol) in MeOH (1 mL) was added to the solution and the stirring was continued for 10 min. The crude mixture was poured into cold hexane (160 mL) and stirred for 1 h. The precipitate was collected by centrifugation and washed by cold hexane for three times. The resulting crude product was extracted with an EtOH:toluene = 1:1 mixture (3 mL) and reprecipitated by cold hexane. The precipitate was collected by centrifugation and extracted with Et_2O (4 mL). The supernatant was precipitated by hexane and dried in vacuo. The yield was 20.7 mg (43% based on the initial Pd). The product was used for the characterization without further crystallization.

Synthesis of ligand-substituted superatom $[PdAu_{24}(C{\equiv}CAr^F)_{18-x}(C{\equiv}CPh)_x](PPh_4)_2$ was conducted by the same procedure as that of $[PdAu_{24}(C{\equiv}CAr^F)_{18}](PPh_4)_2$ except that a mixture of $Ar^FC{\equiv}CH$ and PhC≡CH was used instead of pure $Ar^FC{\equiv}CH$. The yield was 13.1 mg, 10.1 mg, and 6.8 mg with the mixing molar ratio of PhC≡CH of 10, 33, and 50% respectively.

4.2.3 *Photoelectron Spectroscopy*

Anion PES was conducted by using a home-built apparatus equipped with an electrospray ionization (ESI) source, a time-of-flight mass spectrometer (TOF–MS), and a magnetic-bottle type photoelectron spectrometer (MB-PES) described in Chap. 2. $[PdAu_{24}(C{\equiv}CAr^F)_{18-x}(C{\equiv}CPh)_x](PPh_4)_2$ samples dispersed in dichloromethane (~0.2 mg/mL) were introduced into the apparatus via the ESI source (bias voltage = −3 kV, flow rate = 1.5 μL/min) and desolvated by a capillary (O.D. = 1/32 in. and I.D. = 0.5 mm) resistively heated up to 90 °C and biased to −60 V $[PdAu_{24}(C{\equiv}CAr^F)_{18-x}(C{\equiv}CPh)_x]^{2-}$ dianions were transferred to a linear quadrupole ion trap (LQT) through a skimmer (1mm diameter). RF amplitude, frequency, and offset voltage were set to 800 $V_{p\text{-}p}$, 500 kHz, and −15 V, respectively. Continuously flowing ions were accumulated in the LQT for 98 ms with the exit voltage of −25 V and released 2 ms before the acceleration of TOF–MS with the exit voltage of −11 V. Released ions were accelerated up to ~25 keV by a set of Wiley–McLaren type electrodes and temporally and spatially separated by TOF–MS with a mass resolution ($\Delta m/m$) of ~600. For the PES measurement, pulsed third or fourth harmonics of

Nd:YAG laser (Quanta-Ray GCR130, Spectra-Physics, U.S.) were irradiated to the mass-selected ion beam of $[PdAu_{24}(C{\equiv}CAr^{F})_{18-x}(C{\equiv}CPh)_{x}]^{2-}$ during the TOF–MS. Emitted photoelectrons were collected by an inhomogeneous magnetic field generated by a permanent magnet and a coil. The kinetic energies of photoelectrons were determined by TOF of electrons and then converted to the electron binding energies (EBE). The calibration of EBE was done with the PE spectrum of I^{-} produced by the dissociative electron attachment to CH_3I. The energy resolution was about 110 mV for an electron with a kinetic energy of 1 eV. The PE spectra presented in this paper were obtained by accumulating 5000–10,000 laser shots. The obtained PE spectra were smoothed by averaging eleven neighboring data points. The adiabatic electron affinity was determined by the energy of the spectral onset at which the second derivative of the PE spectrum took a local maximum [8].

4.2.4 Theoretical Calculation

Electronic structures of $[PdAu_{24}(C{\equiv}CAr^{F})_{18-x}(C{\equiv}CPh)_{x}]^{2-}$ (**PdAu$_{12}$/*x***$^{2-}$; $x = 0, 1, 2, 6, 18$) were investigated by DFT calculations. DFT calculations were conducted with PBE functionals and def-SV(P) basis sets with def-ecp effective core potentials for Au and Pd by using the Turbomole software package [40].

Initial structures of calculated clusters were constructed as follows:

PdAu$_{12}$/0a^{2-}: A reported crystal structure of $[PdAu_{24}(C{\equiv}CAr^{F})_{18}]^{2-}$ was used.
PdAu$_{12}$/1a^{2-}: A modified structure of **PdAu$_{12}$/0a**$^{2-}$ with one $C{\equiv}CAr^{F}$ away from the $PdAu_{12}$ core replaced by C≡CPh was used.
PdAu$_{12}$/1b^{2-}: A modified structure of **PdAu$_{12}$/0a**$^{2-}$ with one $C{\equiv}CAr^{F}$ bonded to the $PdAu_{12}$ core replaced by C≡CPh was used.
PdAu$_{12}$/2a^{2-}: A modified structure of a more stable isomer **PdAu$_{12}$/1a**$^{2-}$ with one $C{\equiv}CAr^{F}$ away from the C≡CPh ligand replaced by C≡CPh was used.
PdAu$_{12}$/2b^{2-}: A modified structure of a more stable isomer **PdAu$_{12}$/1a**$^{2-}$ with one $C{\equiv}CAr^{F}$ closest to the C≡CPh ligand replaced by C≡CPh was used.
PdAu$_{12}$/6a^{2-}: A modified structure of **PdAu$_{12}$/0a**$^{2-}$ with one $C{\equiv}CAr^{F}$ away from the $PdAu_{12}$ core replaced by C≡CPh for six $Au_2(C{\equiv}CAr^{F})_3$ oligomers was used.
PdAu$_{12}$/18a^{2-}: A modified structure of **PdAu$_{12}$/0a**$^{2-}$ with all $C{\equiv}CAr^{F}$ replaced by eighteen C≡CPh was used.

Structural optimization was conducted for these clusters with fixed spin multiplicity in the singlet state. Although imaginary frequencies were observed in **PdAu$_{12}$/1b**$^{2-}$ and **PdAu$_{12}$/18a**$^{2-}$, the absolute frequencies were small ($<5\ cm^{-1}$) and the structural displacements were due to the combination of the rotation and bending of alkynyl ligands, so further structural optimizations and frequency calculations were not conducted. The vertical detachment energies (VDEs) were calculated by subtracting the energy of the structure-optimized doubly charged anion from that of the singly charged anion with the fixed geometry without considering the zero-point energy.

The PE spectrum of **PdAu$_{12}$/0a**$^{2-}$ was simulated based on the density of states considering the spectral cutoff by the RCB according to the previous report [31]. If the energy of photoelectrons (E_e) obtained by subtracting the binding energy of electrons (E_{BE}) from the photon energy (E_{photon}) is larger than the height of the RCB (E_{RCB}), photoelectrons can leave the cluster without consideration. Even if E_e is smaller than E_{RCB}, photoelectrons can leave the cluster by tunneling with a certain probability. Assuming that the RCB is a rectangular barrier with the height E_{RCB} and the width Δx, the tunneling probability $P(E_e)$ is expressed by the following formula based on the well-established WKB approximation:

$$P(E_e) = \exp\left[-2\sqrt{\frac{2m_e}{\hbar^2}\left(E_{\mathrm{RCB}} + E_{\mathrm{BE}} - E_{\mathrm{photon}}\right)} \times \Delta x\right]. \quad (4.1)$$

The PE spectrum was simulated by assuming that the photoelectron detachment cross section of each energy level was same except for the tunneling probability $P(E_e)$. E_{RCB} and Δx were set to 1.1 eV and 2.5 Å, respectively.

4.3 Results and Discussion

4.3.1 Synthesis and Characterization

The ligand-substituted clusters $[PdAu_{24}(C{\equiv}CAr^F)_{18-x}(C{\equiv}CPh)_x]^{2-}$ (**PdAu$_{12}$/*x***$^{2-}$) were synthesized by the hydride-mediated conversion reaction of $[PdAu_8(PPh_3)_8]^{2+}$ according to our previous report [17]. Synthesis by using a mixture of HC≡CAr^F and HC≡CPh afforded **PdAu$_{12}$/*x***$^{2-}$. ESI mass spectra (ESI–MS, Fig. 4.2a) revealed that the products were mixtures of **PdAu$_{12}$/*x***$^{2-}$ ($x = 0$–6) and that the average value of x could be tuned by the mixing ratio of the two ligands in the synthesis. More than six C≡CPh ligands could not be introduced even by further increasing the mixing ratio of C≡CPh, probably due to intrinsic instability of C≡CPh-rich clusters. UV–visible absorption spectra of **PdAu$_{12}$/*x***$^{2-}$ (Fig. 4.2b) showed similar profiles with distinct peaks at 740 and 535 nm [17] regardless of the x values. These observations indicated that the geometrical structures and the relative energy levels of electronic orbitals were retained in the ligand-substituted superatoms. Differential pulse voltammetry (DPV) of **PdAu$_{12}$/*x***$^{2-}$ dissolved in dichloromethane (Fig. 4.2c) showed that the redox peaks were shifted to a lower voltage as the "average" number of x increased. The linear least square fitting showed that the first oxidation potential linearly decreased by 55.0 ± 0.7 mV per x. This shift clearly indicated that electron-withdrawing CF_3 groups in C≡CAr^F contributed to the stabilization of the 1P HOMOs, while the electronic structure of each composition was unclear in this way due to the broad peak width.

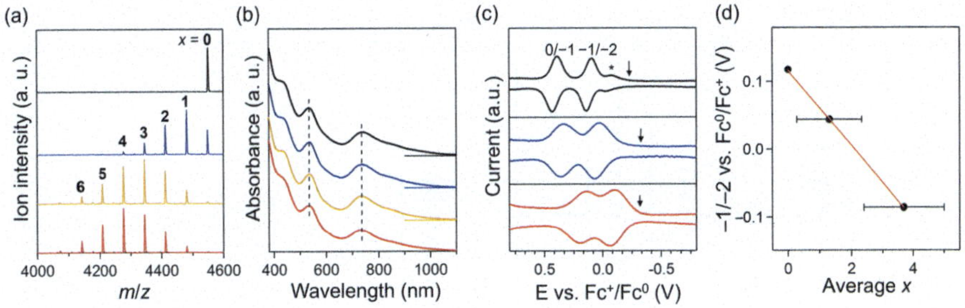

Fig. 4.2 **a** ESI mass spectra, **b** UV–vis absorption spectra, and **c** DPV charts of **PdAu$_{12}$/*x***$^{2-}$ synthesized with different mixing ratios of HC≡CPh and HC≡CAr^F. Black, blue, orange, and red spectra were obtained from the products synthesized with 0, 10, 33, and 50% of HC≡CPh, respectively. The numbers in panel (**a**) correspond to the *x* values. The asterisk and arrows in panel (**c**) show the redox peak of Fc$^+$/Fc^0 and open circuit potentials, respectively. **d** The first oxidation potential of **PdAu$_{12}$/*x***$^{2-}$ in panel (**c**) is plotted against the average number of C≡CPh estimated from the ESI–MS spectra (black) and its fitting line (red). Error bars show the standard deviation of *x* in ESI–MS spectra

4.3.2 Photoelectron Spectroscopy

Gas-phase PES on mass-selected **PdAu$_{12}$/*x***$^{2-}$ was conducted by using a home-built apparatus shown in Chap. 2. **PdAu$_{12}$/*x***$^{2-}$ with a given *x* was isolated temporally and spatially in the TOF–MS, instead of using samples purified by laborious high-performance liquid chromatography (HPLC) [41]. The PE spectra of **PdAu$_{12}$/0**$^{2-}$–**PdAu$_{12}$/6**$^{2-}$ were recorded by adjusting the timing of irradiation of a pulsed output of the Nd:YAG laser. The resolution of the home-built mass spectrometer ($m/\Delta m$ ~ 400) was high enough to distinguish the number of ligand substitution in PES, although it was worse than that of the commercial one ($m/\Delta m$ ~ 5000) (Fig. 4.3).

Figure 4.4a shows PE spectra of [PdAu$_{24}$(C≡CAr^F)$_{18}$]$^{2-}$ (**PdAu$_{12}$/0**$^{2-}$) recorded at 266 nm (4.66 eV) and 213 nm (5.82 eV). PE spectra exhibited three bands at 2.5, 3.8 and 4.7 eV (referred to as the band α, I, and II). The laser-fluence dependence of the PE spectrum of **PdAu$_{12}$/0**$^{2-}$ (Fig. 4.4a, red and black traces) indicated that the band α was due to the two-photon photoelectron detachment process via photoexcited states as discussed in Chap. 3:

$$\mathbf{PdAu_{12}/0}^{2-} + h\nu \rightarrow \mathbf{PdAu_{12}/0}^{2-}*$$

$$\mathbf{PdAu_{12}/0}^{2-} * + h\nu \rightarrow \mathbf{PdAu_{12}/0}^{-} + e^{-}$$

The PE counts of the band I recorded at 266 nm rapidly decreased after 3.6 eV while the band I recorded at 213 nm reached maximum at 3.8 eV (Fig. 4.4b), because the RCB between the detached electron and the remaining anion **PdAu$_{12}$/0**$^-$ (Fig. 4.1b) prevents the detachment of electrons with low kinetic energies as introduced in Sect. 1.3 [28]. The height of the RCB of **0** was estimated to be ~1.1

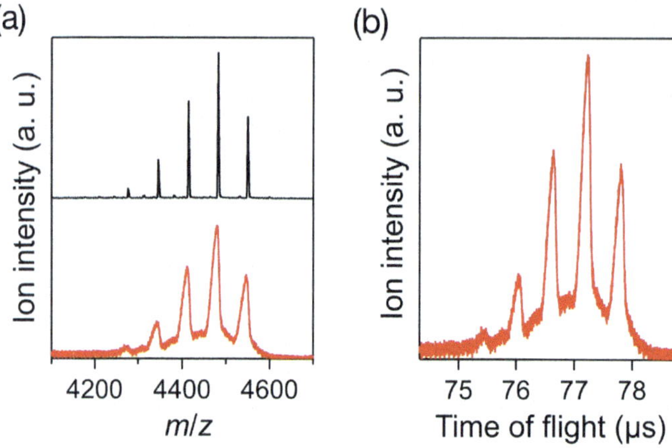

Fig. 4.3 **a** Comparison of mass spectra of **$PdAu_{12}$/*x***$^{2-}$ recorded by a commercial ESI–MS (black) and the home-built ESI–MS (red). **b** Time-of-flight data of mass spectra recorded by the home-built ESI–MS. Broad background signals in panel b are probably due to the artifact of mass spectrometer such as insufficient ion focusing

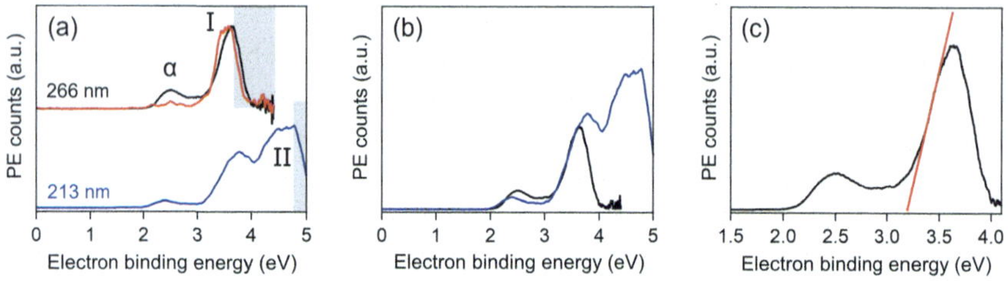

Fig. 4.4 **a** PE spectra of **$PdAu_{12}$/0**$^{2-}$ recorded at 266 nm (top) and 213 nm (bottom). The laser intensity of the black and red trace was 25 and 1.0 mJ cm^{-2}, respectively. **b** Overlapped PE spectra of **$PdAu_{12}$/0**$^{2-}$ recorded at 266 nm (black) and 213 nm (blue). **c** The PE spectrum of **$PdAu_{12}$/0**$^{2-}$ recorded at 266 nm (black) and the tangent line at the inflection point of the band I (red)

eV from the difference between the photon energy (4.66 eV) and the critical EBE of 3.6 eV. This RCB height was comparable to that (~1.1 eV) of $[MAg_{24}(SR)_{18}]^{2-}$ (M = Pd, Pt) discussed in Chap. 3. The adiabatic electron affinity (AEA) of the corresponding monoanion **$PdAu_{12}$/0**$^{-}$ was determined to be 3.19 ± 0.06 eV based on the crossing point of the tangent line at the inflection point of the band I and the horizontal axis (Fig. 4.4c).

To help the assignment of the PE spectra, DFT calculations were conducted on **$PdAu_{12}$/0**$^{2-}$ by using the full ligands. Figure 4.5a shows the shape of Kohn–Sham orbitals of the optimized structure (referred to as **$PdAu_{12}$/0a**$^{2-}$) using the Turbomole program [40]. Based on the shape of orbitals, HOMO, HOMO–1 and HOMO–2 were assigned to superatomic 1P orbitals and C(sp) while HOMO–3 and deeper ones were assigned to Au(5d) and C(sp). The PE spectrum was simulated based on density of states of **$PdAu_{12}$/0a**$^{2-}$ with the RCB effect included by assuming that slow electrons

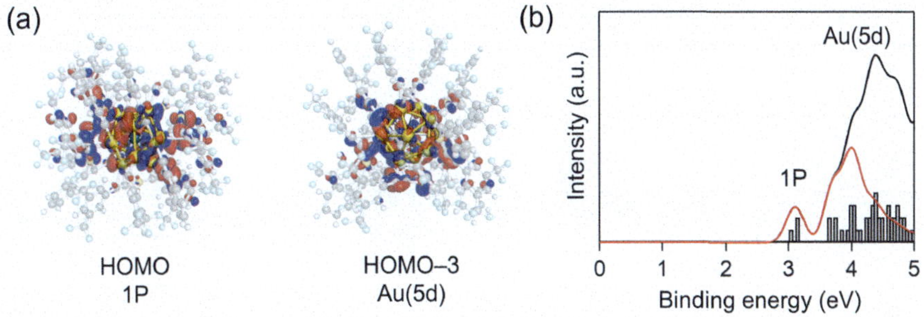

Fig. 4.5 **a** The shape of HOMO and HOMO−3 of **PdAu$_{12}$/0a^{2-}**. A contour value was set to 0.015. **b** The simulated PE spectrum of **PdAu$_{12}$/0a^{2-}** without (black line) and with (red line) considering the spectral cutoff by a repulsive Coulomb barrier. Density of states was made by broadening each energy level with full-width half maximum of 0.3 eV. The histogram (bars) showed the population of the energy levels with a bin width of 50 meV. To compensate the translational energy difference between experimental and theoretical spectra, photon energy was increased by 0.7 eV by adding the difference of the energy of the first peak of the experimental and simulated PE spectra

can leave the dianions via tunneling with some probability (Sect. 4.2.4, Fig. 4.5b). The simulated PE spectrum reproduced the main features of the 213 nm spectrum and the band I and II were assigned to superatomic 1P orbitals and Au(d) + C(sp) bands, respectively.

PES of ligand-substituted superatoms **PdAu$_{12}$/1^{2-}–PdAu$_{12}$/6^{2-}** was conducted by using the high-fluence laser to maintain the S/N ratio of the PE spectra. PE spectra recorded at 266 nm are superimposed in Fig. 4.6a after normalization of the height of the band I. The PE spectra were clearly red-shifted to lower EBE by increasing the number of C≡CPh (x) while keeping the spectral profiles unchanged. The band α due to the two-photon PE detachment process via photoexcited states was commonly observed for **PdAu$_{12}$/1^{2-}–PdAu$_{12}$/6^{2-}**, indicating that the lifetime of the excited state of **PdAu$_{12}$/1^{2-}–PdAu$_{12}$/6^{2-}** was in the same order with **PdAu$_{12}$/0^{2-}**. The height of RCB determined from the apparent peak position of the band I at 266 nm was gradually shifted to the lower EBE, showing that the RCB becomes higher with increase in x. This finding revealed that the RCB height was determined not only by the charge state of the cluster but also the nature of the ligand. The AEAs of the corresponding monoanions of **PdAu$_{12}$/1^{-}–PdAu$_{12}$/6^{-}** were determined based on the 266 nm spectra in the same manner with **PdAu$_{12}$/0^{-}** (Fig. 4.4c). Figure 4.6b plots the AEAs thus determined as a function of x and indicates that both AEAs decrease linearly with respect to x. The amounts of the decrease of AEAs by the exchange of one C≡CAr^F with one C≡CPh was calculated to be 74 ± 3 meV, by the linear least squares fitting of the plot. Linear extrapolation of the plot to $x = 18$ predicts that the AEA of monoanion of a putative cluster [PdAu$_{24}$(C≡CPh)$_{18}$]$^-$ to be 1.86 eV. This estimation suggests that the 1P HOMOs of [PdAu$_{24}$(C≡CAr^F)$_{18}$]$^{2-}$ are more stable than those of [PdAu$_{24}$(C≡CPh)$_{18}$]$^{2-}$ by 1.3 eV. A consistent behavior was observed by differential pulse voltammetry (DPV) of **PdAu$_{12}$/x^{2-}** dissolved in dichloromethane (Fig. 4.2). It should be noted that the steric effect of ligands

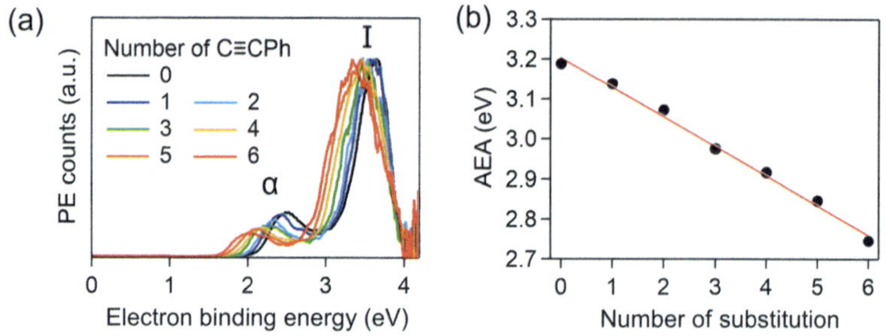

Fig. 4.6 **a** The PE spectra of **$PdAu_{12}/0^{2-}$–$PdAu_{12}/6^{2-}$**. **b** Experimental values AEAs (circles) of **$PdAu_{12}/0^{2-}$–$PdAu_{12}/6^{2-}$** and the fitting line plotted against the number of substituted C≡CPh ligands

also plays a vital role in the stability and structure of ligand-protected gold clusters [42]. The previous study showed that $[PdAu_{24}(C{\equiv}CR)_{18}]^{2-}$ can be synthesized with R = 3,5-$(CF_3)_2C_6H_3$ and 3-$NO_2C_6H_4$, while not with R = Ph, 4-$(CH_3)_3CC_6H_4$, 4-$CF_3C_6H_4$, 3-$CH_3C_6H_4$, 3-$CH_3OC_6H_4$, PhC_2H_4, and 2-SC_2H_4Ph[17]. The specificity suggested that both steric and electronic effect were key factors to generate $[PdAu_{24}(C{\equiv}CR)_{18}]^{2-}$. Nevertheless, we can conclude that the electron-withdrawing CF_3 substituents in C≡CAr^F significantly contribute to the electronic stabilization of $[PdAu_{24}(C{\equiv}CAr^F)_{18}]^{2-}$.

4.3.3 DFT Calculation

To elucidate the origin of the substituent effects on electronic structures, DFT calculations were also conducted on **$PdAu_{12}/x^{2-}$** (x = 0, 1, 2, 6, and 18) by using the full ligands. Figure 4.7 shows geometric structures optimized by using the Turbomole program [40]. The effects of the location of C≡CPh ligand(s) on the electronic structures were studied for $x = 1$ and 2. The VDEs and energies of the 1P HOMOs agreed within ± 0.01 eV regardless of the location of C≡CPh ligands (Table 4.1). This result indicates that the PE spectra and AEAs in Fig. 4.6 are not significantly affected by the spatial distribution of the two ligands. Figure 4.8a plots the calculated VDE values of **$PdAu_{12}/x^{2-}$** averaged over the regio-isomers. The averaged VDEs decreased linearly with respect to x by 89 ± 1 meV per one ligand exchange. To clarify the origin of the decrease in the VDE upon ligand exchange, the density of states of **$PdAu_{12}/x^{2-}$** are examined in Fig. 4.9. The 1P HOMOs are upshifted from −2.07 to −0.70 eV ongoing from $x = 0$ to 18. Since the energy difference between the 1P HOMOs of **$PdAu_{12}/0a^{2-}$** and **$PdAu_{12}/18a^{2-}$** (~1.4 eV) qualitatively agrees with the decrease of the VDEs (1.5 eV), I conclude that the destabilization of the 1P HOMOs by the ligand exchange is the direct cause of the decrease in the VDEs. The natural charges obtained by natural population analysis showed that the exchange of one C≡CAr^F ligand with one C≡CPh ligand induces a negative charge of − 0.24e at the C_6 ring in the phenyl moiety (Fig. 4.8b, Table 4.2) while the charge at the $PdAu_{12}$ core was barely changed (−0.004e). Even after the complete ligand exchange, the natural charge of the $PdAu_{12}$ core of **$PdAu_{12}$/18a**$^{2-}$ was only slightly negative by −0.07e compared with **$PdAu_{12}$/0a**$^{2-}$. A similar behavior was reported in the electrochemical redox potentials of $[Au_{25}(SC_2H_4Ph)_{18-x}(SC_6H_4Br)_x]^-$ and the electronic charge distribution of $[Au_{25}(SCH_3)_{18-x}(SCH_2Cl)_x]^-$ with an increase in the number of polar thiols (x) [20]. Therefore, the observed ligand effect on the electronic stabilization is mainly ascribed to the inductive mechanism through space, rather than the resonance mechanism mediated by π electrons [15, 16].

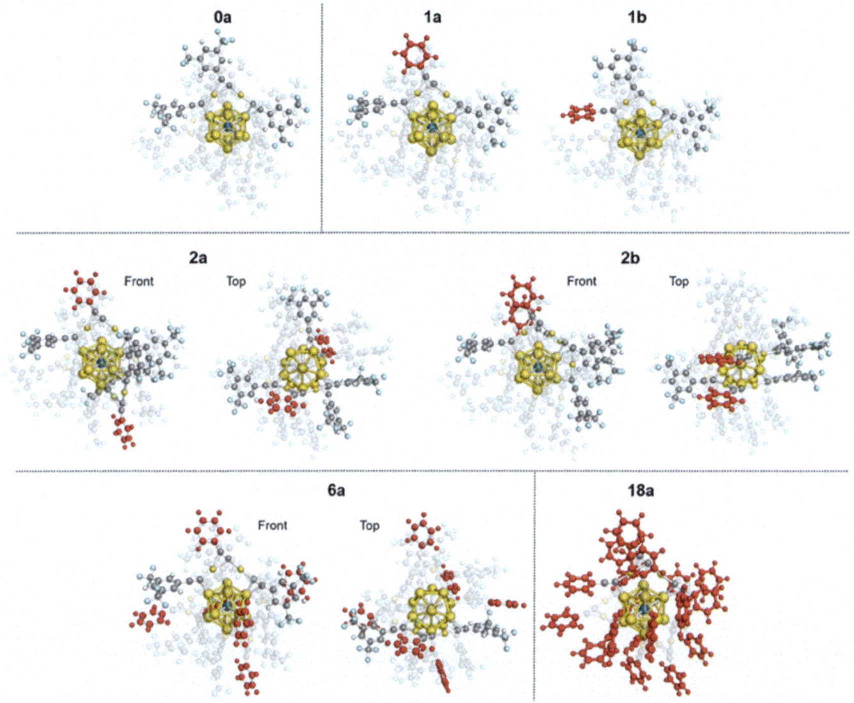

Fig. 4.7 Geometric structures of DFT-optimized **$PdAu_{12}$/x**$^{2-}$ (x = 0, 1, 2, 6, 18). Color code: yellow = Au; blue = Pd, gray = C, pale blue = F, white = H. Ph groups are colored in red for clarity

Table 4.1 Total energies, energies of HOMO, and vertical detachment energies (VDEs) of position isomers **$PdAu_{12}$/1a**$^{2-}$, **$PdAu_{12}$/1b**$^{2-}$, **$PdAu_{12}$/2a**$^{2-}$, and **$PdAu_{12}$/2b**$^{2-}$

	ΔE_{total} (eV)	E_{HOMO} (eV)	VDE (eV)
$PdAu_{12}$/1a^{2-}	0	−1.99	2.9
$PdAu_{12}$/1b^{2-}	0.016	−1.98	2.89
$PdAu_{12}$/2a^{2-}	0	−1.92	2.83
$PdAu_{12}$/2b^{2-}	0.06	−1.93	2.84

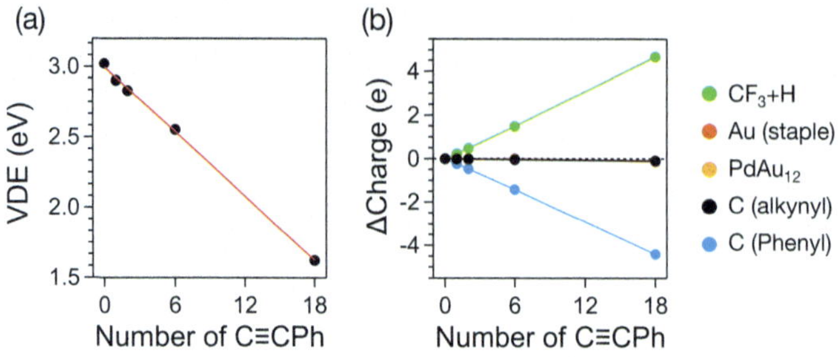

Fig. 4.8 **a** Calculated VDEs of **PdAu$_{12}$/*x***$^{2-}$ (black dots) and the fitting line (red). **b** Evolution of the natural charges obtained by natural population analysis with respect to that of **PdAu$_{12}$/0a**$^{2-}$

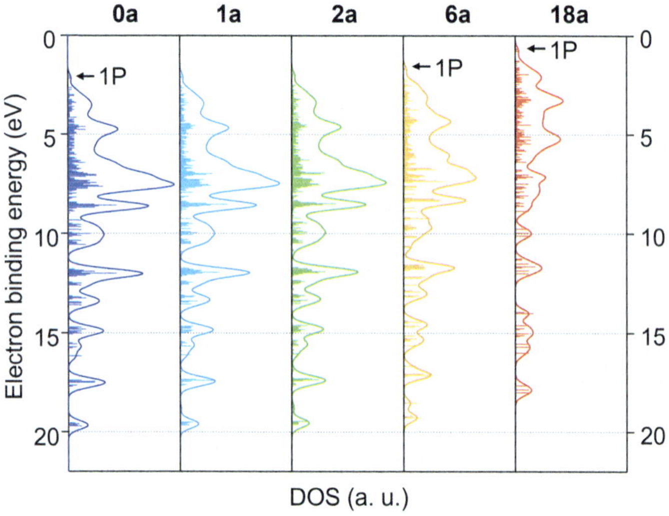

Fig. 4.9 Density of states of **PdAu$_{12}$/*x***$^{2-}$ ($x = 0, 1, 2, 6, 18$)

Table 4.2 Natural charges obtained by natural population analysis of **PdAu$_{12}$/*x***$^{2-}$

	$CF_3 + H$	Au (staple)	$PdAu_{12}$	C (alkynyl)	C (phenyl)
PdAu$_{12}$/0a^{2-}	15.81	6.58	2.91	−10.71	−16.59
PdAu$_{12}$/1a^{2-}	16.06	6.56	2.92	−10.72	−16.82
PdAu$_{12}$/1b^{2-}	16.06	6.58	2.91	−10.72	−16.82
PdAu$_{12}$/2a^{2-}	16.31	6.55	2.92	−10.72	−17.05
PdAu$_{12}$/2b^{2-}	16.31	6.56	2.92	−10.73	−17.07
PdAu$_{12}$/6a^{2-}	17.32	6.48	2.95	−10.75	−18.80
PdAu$_{12}$/18a^{2-}	20.50	6.46	2.84	−10.81	−20.99

4.3.4 Effect of Ligand Substitution

The modulation of the EBEs of Pd@Au_{12} superatom by changing the nature of the alkynyl ligands is phenomenologically similar to the manipulation of the work function of the bulk gold ($W_{Au} = 4.9$ eV) by forming the self-assembled monolayers (SAMs) of thiols [43–45]. For example, the SAM of perfluorinated $C_8F_{17}C_2H_4S$ (dipole moment normal to the surface $\mu_\perp = -2.19$ D) raised W_{Au} by 0.6 eV, whereas that of n-$C_{16}H_{33}S$ ($\mu_\perp = 1.60$ D) decreased W_{Au} by 0.8 eV [43]. This behavior was explained by the downshift and upshift of the vacuum level (E_{vac}), respectively, with respect to the Fermi level of bulk gold (F_{Au}) by the interface dipole [43]. Considering the metal surface covered by thiolates (Fig. 4.10a), the change of work function compared to the pristine metal is calculated to be

$$\Delta W = Ne\left(\frac{\mu_\perp}{\varepsilon} - \mu_{Au-S}\right) \tag{4.1}$$

where N, $\mu_{\perp,mol}$, μ_{Au-S}, and ε is the surface density of molecules, the dipole moment of molecules projected to the axis normal to the surface, the intrinsic dipole moment of Au–S bond, and the dielectric constant of molecules, respectively [43].

In contrast, the ligand-protected superatoms can be modeled by a metallic sphere covered by an electrical double layer (Fig. 4.10b). Assuming a neutral metallic sphere with a radius of R covered by an electrical double layer with a charge of $\pm\delta e$ and a thickness of D (Fig. 4.10b). According to the Gauss's law of spherical charge distribution, the electric field E as the function of radial coordinate r is described as

$$E(r) = \frac{Q}{4\pi\varepsilon_0 r^2} \tag{4.2}$$

where Q and ε_0 is the total charge inside the sphere with the radius of r and the permittivity of vacuum, respectively. The electric field in case of Fig. 4.10b is written as follows:

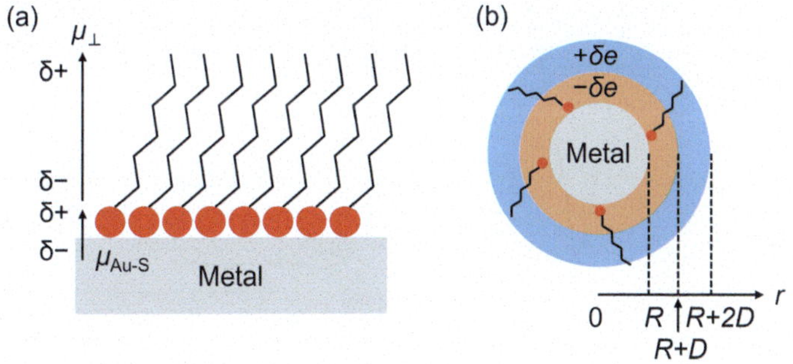

Fig. 4.10 Schematic illustration of the simplified model of **a** a metal surface and **b** a metal sphere covered by organic molecules

$$r < R \text{ or } R + 2D < r : E(r) = 0 \tag{4.3}$$

$$R \le r \le R + D : E(r) = \frac{-\delta e}{4\pi \varepsilon_0 r^2} \times \frac{r^3 - R^3}{(R+D)^3 - R^3} \tag{4.4}$$

$$R + D \le r \le R + 2D : E(r) = \frac{-\delta e}{4\pi \varepsilon_0 r^2} \times \left(1 - \frac{r^3 - (R+D)^3}{(R+2D)^3 - (R+D)^3}\right) \tag{4.5}$$

It is noteworthy that the electric field is pointing inward if the inner layer is negatively charged: a positive charge is attracted while a negative charge is repelled. Assuming the potential at infinity $V(r = \infty)$ is equal to zero, the potential at the metallic sphere is obtained by integrating $E(r)$ from infinity to $r = R$. The example of the electric field and the electric potential is shown in Fig. 4.11 with $R = 10$Å, $D = 5$Å, and $\delta e = 1e$. It is noteworthy that if the inner layer is negatively charged, (1) the electric potential at the metallic core was qualitatively negative and uncomfortable for electrons (2) the electric potential was proportional to δe. The simple model calculation indicated that the electrostatic effect of the dipole moment of ligands influences the electric potential of the metallic core, and the degree is proportional to the dipole moment.

Based on the above discussion, the current observation of the ligand effect in **$PdAu_{12}/x^{2-}$** is illustrated in Fig. 4.12. the 1P orbitals of **$PdAu_{12}/0^{2-}$** can be destabilized by electrostatic interaction with the permanent dipoles of $Au_2(C{\equiv}CAr^F)_3$ pointing to the Pd@Au_{12} core: $\mu = 0.74$ D for C≡CAr^F (Fig. 4.10). By replacing some of the C≡CAr^F ligands with C≡CPh ($\mu = 2.00$ D), the nominal dipole moments of the surface layer increase, leading to further destabilization of the 1P orbitals. As a result, the EBE of **$PdAu_{12}/x^{2-}$** is reduced with increase in x. This study showed that the simple electrostatic model developed for the surface modification of bulk metals was applicable to the quantized small metal clusters.

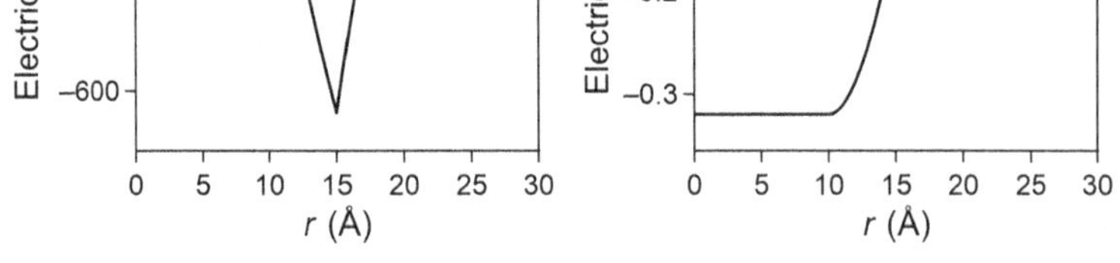

Fig. 4.11 **a** Electric field and **b** electric potential made by a metal sphere covered by an electrical double layer. R, D, and δe were set to 10 Å, 5 Å, and 1e, respectively

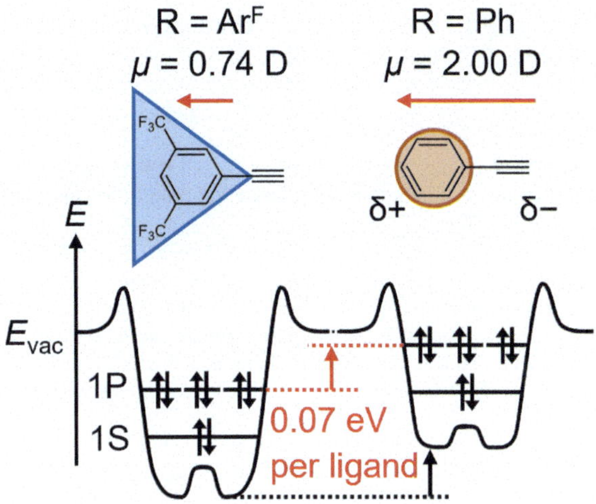

Fig. 4.12 Schematic illustration of the effect of ligand substitution on the jellium potential of eight-electron superatoms

4.4 Summary

In summary, we have clarified the vital role of electron-withdrawing substituents on the electronic stability of alkynyl-protected structurally defined superatoms by anion PES on ligand-substituted superatoms $[PdAu_{24}(C{\equiv}CAr^F)_{18-x}(C{\equiv}CPh)_x]^{2-}$. PES of mass-selected clusters showed that the EBEs of the clusters decreased linearly by replacing $C{\equiv}CAr^F$ with $C{\equiv}CPh$, indicating that electron-withdrawing CF_3 substituents electronically stabilize the clusters. DFT calculations on the ligand-substituted clusters reproduced the decrease of the EBEs with increase in x and the ligand effect was explained by an electrostatic mechanism similar to the modulation of the work function of Au film by organic monolayers. This study has not only revealed the hidden role of electron-withdrawing substituents on the electronic stability of alkynyl-protected Au clusters but also provides a new guiding principle for tuning the electronic structures by ligand exchange for their applications in electrocatalysts and electrochemical sensing [46].

References

1. Tsukuda T, Häkkinen H (2015) Protected metal clusters: from fundamentals to applications, 1st edn. Amsterdam, The Netherlands, Elsevier B.V.
2. Jin R, Zeng C, Zhou M, Chen Y (2016) Chem Rev 116:10346
3. Wan X-K, Guan Z-J, Wang Q-M (2017) Angew Chem Int Ed 56:11494
4. Lei Z, Li J-J, Wan X-K, Zhang W-H, Wang Q-M (2018) Angew Chem Int Ed 57:8639
5. Li J-J, Guan Z-J, Lei Z, Hu F, Wang Q-M (2019) Angew Chem Int Ed 58:1083
6. Ito S, Takano S, Tsukuda T (2019) J Phys Chem Lett 10:6892
7. Han X-S, Luan X, Su H-F, Li J-J, Yuan S-F, Lei Z, Pei Y, Wang Q-M (2020) Angew Chem Int Ed 59:2309
8. Guan Z-J, Hu F, Li J-J, Liu Z-R, Wang Q-M (2020) Nanoscale 12:13346

9. Guan Z, Hu F, Li J, Wen Z, Lin Y, Wang Q (2020) J Am Chem Soc 142:2995
10. Wang J-Q, Shi S, He R, Yuan S, Yang G, Liang G, Wang Q (2020) J Am Chem Soc 142:18086
11. Lei Z, Wan X-K, Yuan S-F, Guan Z-J, Wang Q-M (2018) Acc Chem Res 51:2465
12. Wang Y, Wan X, Ren L, Su H, Li G, Malola S, Lin S, Tang Z, Häkkinen H, Teo BK et al (2016) J Am Chem Soc 138:3278
13. Wan X, Wang J-Q, Nan Z, Wang Q (2017) Sci Adv 3:e1701823
14. Kawawaki T, Ebina A, Hosokawa Y, Ozaki S, Suzuki D, Hossain S, Negishi Y (2021) Small 17:2005328
15. Kobayashi N, Kamei Y, Shichibu Y, Konishi K (2013) J Am Chem Soc 135:16078
16. Sugiuchi M, Shichibu Y, Nakanishi T, Hasegawa Y, Konishi K (2015) Chem Commun 51:13519
17. Takano S, Ito S, Tsukuda T (2019) J Am Chem Soc 141:15994
18. Kang X, Chong H, Zhu M (2018) Nanoscale 10:10758
19. Quo R, Murray RW (2005) J Am Chem Soc 127:12140
20. Parker JF, Kacprzak KA, Lopez-Acevedo O, Häkkinen H, Murray RW (2010) J Phys Chem C 114:8276
21. Aikens CM (2010) J Phys Chem Lett 1:2594
22. Tlahuice-Flores A, Whetten RL, Jose-Yacaman M (2013) J Phys Chem C 117:20867
23. Ito S, Koyasu K, Takano S, Tsukuda T (2020) J Phys Chem C 124:19119
24. Koga N, Morokuma K (1985) Chem Phys Lett 119:371
25. Zhang SB, Cohen ML, Chou MY (1987) Phys Rev B 36:3455
26. Hirai H, Ito S, Takano S, Koyasu K, Tsukuda T (2020) Chem Sci 11:12233
27. Wang L-S, Wang X-B (2000) J Phys Chem A 104:1978
28. Hirata K, Yamashita K, Muramatsu S, Takano S, Ohshimo K, Azuma T, Nakanishi R, Nagata T, Yamazoe S, Koyasu K et al (2017) Nanoscale 9:13409
29. Kim K, Hirata K, Nakamura K, Kitazawa H, Hayashi S, Koyasu K, Tsukuda T (2019) Angew Chem Int Ed 58:11637
30. Hirata K, Kim K, Nakamura K, Kitazawa H, Hayashi S, Koyasu K, Tsukuda T (2019) J Phys Chem C 123:13174
31. Tasaka Y, Nakamura K, Malola S, Hirata K, Kim K, Koyasu K, Häkkinen H, Tsukuda T (2020) J Phys Chem Lett 11:3069
32. Veenstra AP, Monzel L, Baksi A, Czekner J, Lebedkin S, Schneider EK, Pradeep T, Unterreiner A-N, Kappes MM (2020) J Phys Chem Lett 11:2675
33. Hirata K, Tomihara R, Kim K, Koyasu K, Tsukuda T (2019) Phys Chem Chem Phys 21:17463
34. Chakraborty P, Pradeep T (2019) NPG Asia Mater. 11:48
35. Cirri A, Hernández HM, Johnson CJ (2020) J Phys Chem A 124:1467
36. Koyasu K, Tsukuda T (2021) J Chem Phys 154:140901
37. Brandys MC, Jennings MC, Puddephatt RJ (2000) J Chem Soc Dalt Trans 4601
38. Takano S, Hirai H, Muramatsu S, Tsukuda T (2018) J Am Chem Soc 140:8380
39. Takano S, Hirai H, Muramatsu S, Tsukuda T (2018) J Am Chem Soc 140:12314
40. TURBOMOLE V7.4; University of Karlsruhe and Forschungs-zentrum Karlsruhe GmbH (1989–2007) and TURBOMOLE GmbH (since 2007); 2019; available from http://www.turbomole.com
41. Negishi Y, Hashimoto S, Ebina A, Hamada K, Hossain S, Kawawaki T (2020) Nanoscale 12:8017
42. Zhang B, Chen J, Cao Y, Chai OJH, Xie J (2021) Small 17:2004381
43. Evans SD, Urankar E, Ulman A, Ferris N (1991) J Am Chem Soc 113:4121
44. Alloway DM, Hofmann M, Smith DL, Gruhn NE, Graham AL, Colorado R, Wysocki VH, Lee TR, Lee PA, Armstrong NR (2003) J Phys Chem B 107:11690
45. Boer DB, Hadipour A, Mandoc MM, Woudenbergh VT, Blom PWM (2005) Adv Mater 17:621
46. Kwak K, Lee D (2019) Acc Chem Res 52:12
47. Maity P, Tsunoyama H, Yamauchi M, Xie S, Tsukuda T (2011) J Am Chem Soc 133:20123

Chapter 5
Bonding Interaction in Homonuclear Superatomic Molecules

Abstract The fifth chapter describes the bonding interaction in chemically synthesized superatomic molecules $[M_1Au_{37}(PET)_{24}]^-$ (PET = 2-PhC_2H_4S; M = Pd, Pt). $Au_{38}(PET)_{24}$ is known to have a bi-icosahedral Au_{23} core consisting of two Au_{13} icosahedrons by sharing three Au atoms. Previous theoretical studies based on a super valence bond (SVB) model have demonstrated that the bonding scheme in the Au_{23} core is similar to that in the F_2 molecule. The SVB model predicted that the electron configuration of the Au_{23} core with 14 valence electrons is expressed as $(1\Sigma)^2(1\Sigma^*)^2(1\Pi)^4(2\Sigma)^2(1\Pi^*)^4$ where each orbital is created by the bonding and antibonding interactions between the 1S and 1P superatomic orbitals of the icosahedral Au_{13} units. Therefore, the bi-icosahedral Au_{23} can be viewed as a di-superatomic molecule. To validate the SVB model, I conducted anion photoelectron spectroscopy (PES) on $[M_1Au_{37}(PET)_{24}]^-$ (M = Pd and Pt), which are isoelectronic and isostructural with $Au_{38}(PET)_{24}$. I conducted density functional theory calculations to confirm the isoelectronicity of $[M_1Au_{37}(PET)_{24}]^-$ and $Au_{38}(PET)_{24}$. Anion PES on $[M_1Au_{37}(PET)_{24}]^-$ observed two distinctive peaks as predicted by the SVB model: one from the nearly degenerate $1\Pi^*$ orbitals and the other from the nearly degenerated 1Π and 2Σ orbitals, providing the experimental support of the SVB model.

Keywords Superatomic molecule · Super valence bond · Molecular orbital · Gold cluster

5.1 Introduction

In Chaps. 3 and 4, I studied eight-electron superatoms with isotropic and spherical icosahedral cores. On the other hand, since pioneering works by Teo [1], an increasing number of ligand-protected gold clusters with anisotropic structures have recently been reported [2–14]. Single crystal X-ray diffraction (SCXRD) analysis revealed that their Au cores are composed of small Au cluster units that are called superatoms based on the similarity of their quantized electronic structures compared with those

S. Ito, *Electronic Structures and Bonding Interaction of Structurally Defined Gold/Silver Superatoms*, Springer Theses, https://doi.org/10.1007/978-981-96-2892-6_5

of conventional atoms. Table 5.1 lists typical examples of core structures of ligand-protected anisotropic Au clusters composed of a variety of superatomic units and bonding patterns. For instance, different types of dimer and trimer of the icosahedral Au_{13} units were formed by sharing a facet [3] or vertex atom(s) [2, 5, 7, 9]. Hydrides mediate bonding between the facets of the Au_{12} units and those of Au_{11} units to form Au_{24} and Au_{22}, respectively [11, 12]. The hydride-mediated bonding between Au_9 and Au_{11} was also found in the Au_{20} heterodimer [10]. Because the structures of the superatomic units are retained, these anisotropic clusters can be viewed as "superatomic molecules" [15–22].

The bonding theory of superatomic molecules is an interesting subject based on an analogy of the chemical bonding theory of conventional molecules [23]. A super valence bond (SVB) model has been proposed for the bi-icosahedral Au_{23} core (the top row of Table 5.1) of a thiolate-protected Au cluster $Au_{38}(PET)_{24}$ (referred to as $\mathbf{Au_{23}}^0$ hereafter, PET = SC_2H_4Ph) as shown in Fig. 1.7 [16, 17, 22]. According to the SVB model, the electron configuration of $\mathbf{Au_{23}}^0$ with 14 electrons is expressed as $(1\Sigma)^2(1\Sigma^*)^2(1\Pi)^4(2\Sigma)^2(1\Pi^*)^4$: the 1Σ and $1\Sigma^*$ orbitals are constructed by the bonding and antibonding interactions between the 1S superatomic orbitals (SOs) of the Au_{13} units, respectively; the $1\Pi/1\Pi^*$ and $2\Sigma/2\Sigma^*$ orbitals were constructed by the

Table 5.1 Examples of Au superatomic molecules[a]

Superatom	Bonding pattern	Superatomic molecules	References
Au_{13}	Face shared	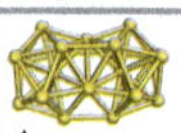 Au_{23}	[3]
	Vertex shared	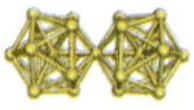 Au_{25}	[2, 9]
		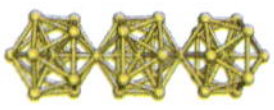 Au_{37}	[5]
Au_{12}	Hydride bridged	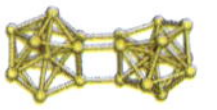 Au_{24}	[12]
Au_{11}	Hydride bridged	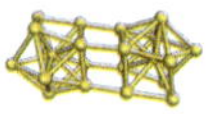 Au_{22}	[11]
Au_9, Au_{11}	Hydride bridged	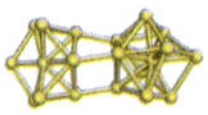 Au_{20}	[10]

[a]The ligand layer is omitted for simplicity

tangential and coaxial overlaps of the 1P SOs of the Au_{13} units [16, 17]. A comparison with conventional diatomic molecules showed that such a bonding scheme in $\mathbf{Au_{23}}^0$ is similar to that in homonuclear F_2 molecules, indicating that $\mathbf{Au_{23}}^0$ can be categorized as homonuclear superatomic molecules. The extension of this idea to $Pd_2Au_{36}(PET)_{24}$ ($\mathbf{Pd_2Au_{21}}^0$) with 12 electrons suggests that the ground state is triplet, similar to O_2 molecules. However, the superconducting quantum interference device (SQUID) magnetometer measurement demonstrated that $\mathbf{Pd_2Au_{21}}^0$ was not a triplet species, but a singlet species [24]. This phenomenon was explained by the breaking of the degeneracy of the highest occupied molecular orbitals (HOMOs) by lowering the core symmetry, implying that a Hund's-type rule is not applicable for $\mathbf{Pd_2Au_{21}}^0$. It was also found that the 1P SOs of the Pd@Au_{12} units in $\mathbf{Pd_2Au_{21}}^0$ overlap in a tilted configuration [24]. Thus, the bonding theory of superatomic molecules is still in its infancy.

A promising experimental approach for validation of the SVB model is gas-phase anion photoelectron spectroscopy (PES) that allows us to probe the absolute energy levels and the density of states of the occupied SOs with reference to the vacuum level [25–28]. A monoanion of $\mathbf{Au_{23}}^0$ is an ideal target for anion PES because it allows mass selection prior to the PE measurement. However, a stable monoanion of $\mathbf{Au_{23}}^0$ could not be produced even with the strong reducing agent $NaBH_4$ because of the closed electronic structure of $\mathbf{Au_{23}}^0$. To circumvent this problem, $[Pd_1Au_{37}(PET)_{24}]^-$ ($\mathbf{Pd_1Au_{22}}^-$) and $[Pt_1Au_{37}(PET)_{24}]^-$ ($\mathbf{Pt_1Au_{22}}^-$) obtained by reducing the corresponding neutrals $Pd_1Au_{37}(PET)_{24}$ ($\mathbf{Pd_1Au_{22}}^0$) and $Pt_1Au_{37}(PET)_{24}$ ($\mathbf{Pt_1Au_{22}}^0$) were utilized for anion PES.

The synthesis and characterization of $\mathbf{M_1Au_{22}}^0$ and $\mathbf{M_1Au_{22}}^-$ was conducted by Ms. Emi Ito, a co-author of the paper related to this chapter. The formation of bi-icosahedral M_1Au_{22} cores with open electronic structure in $\mathbf{M_1Au_{22}}^0$ was confirmed by SCXRD analysis, electron paramagnetic resonance (EPR) measurement, and optical spectroscopy. My contribution is the density functional theory (DFT) calculation and anion PES of $\mathbf{M_1Au_{22}}^-$. The isoelectronicity of $\mathbf{M_1Au_{22}}^-$ (M = Pd, Pt) and undoped $\mathbf{Au_{23}}^0$ was confirmed by DFT calculations. The PE spectra of the mass-selected $\mathbf{M_1Au_{22}}^-$ (M = Pd, Pt) were recorded in the gas phase. Two peaks in the PE spectra were assigned to the PE detachment from the 1Π, 2Σ, and 1Π* orbitals with the help of DFT calculation. In this doctoral thesis, I briefly summarized the synthesis and characterization conducted by Ms. Emi Ito and discussed the results of DFT calculation and anion PES in detail.

5.2 Methods

5.2.1 General

All commercially available chemicals were used without further purification. Sodium borohydride ($NaBH_4$), trimethylamine (NEt_3), acetone, ethanol (EtOH), methanol (MeOH), dichloromethane (DCM), hexane, chloroform, and chlorobenzene were obtained from FujiFilm Wako Pure Chemical Corporation. 2-Phenylethanethiol (PET-H) and *trans*-2-[3-(4-*tert*-butylphenyl)-2-methyl-2-propenylidene]malononitrile (DCTB) were purchased from Tokyo Chemical Industry. Tetrahydrofuran (THF; dehydrated and stabilizer free) and DCM (dehydrated) were purchased from Kanto Chemical Company. 1, 2-Dichlorobenzene was obtained from Merck. $AuClSMe_2$ complex, $MAu_{24}(PET)_{18}$ (M = Pd, Pt), and $[Au_9(PPh_3)_8](NO_3)_3$ were synthesized according to the reported procedures, respectively [29–32].

5.2.2 Synthesis

All the synthesis was conducted by Ms. Emi Ito.

$[Pd_1Au_{37}(SC_2H_4Ph)_{24}]^0$ (**$Pd_1Au_{22}{}^0$**).

A vial was charged with $Pd_1Au_{24}(PET)_{18}$ (9.1 mg, 1.25 μmol), followed by DCM (1.1 mL). Ethanolic solution containing 1 equation of $NaBH_4$ (0.1 M, 12.5 μL) was added to the vial (solution A). One equation of $[Au_9(PPh_3)_8](NO_3)_3$ (5.0 mg, 1.25 μmol) was dissolved in DCM (275 μL) in another vial, followed by the rapid addition of ethanolic solution containing 1 equation of $NaBH_4$ (0.1 M, 12.5 μL) (solution B). Solution B was added dropwise to solution A and the mixture was stirred at room temperature for 20 min. Twelve equation of $AuClSMe_2$ (4.4 mg, 15 μmol) was added to THF (250 μL) in another vial to obtain a white suspension. 6 equation of PET (1.01 μL, 7.5 μmol) was added dropwise to the suspension while stirring to obtain a pale yellow solution (solution C). Solution C was added dropwise to the mixture of solutions A and B and the mixture was stirred at room temperature for more than 130 min. Then, 12 equation of NEt_3 (2.08 μL, 15 μmol) was added to the solution and the mixture was stirred for 10 min. Finally, MeOH was added and the solution was evaporated to dryness. The residue was washed with EtOH and extracted with DCM. The extract was evaporated and then purified by running Al_2O_3 column chromatography. DCM and *n*-hexane mixed in different volume ratios were used as the mobile phase. Firstly, a column was built and the crude product was loaded with the least polar solvent (DCM:*n*-hexane = 1:1). Then the eluent was changed to a more polar solvent (DCM:*n*-hexane = 2:1). In the end, the most polar eluent (DCM:*n*-hexane = 6:1) was used. Green bands eluted first and then the color of the bands gradually changed to brown. The first few fractions turning to brown were discarded because their UV–Vis spectra suggested the contamination by a small amount of $\mathbf{Au_{23}}^0$. The latter brown part of the bands was collected and further purified with PTLC. PTLC was developed with the mixture of DCM and *n*-hexane to obtain green and brown bands. A brown band was collected and extracted with a volume ratio of DCM:MeOH = 50:1. The brown bands were collected and dried *in vacuo*. Yield: 2.3 mg (17% with respect to Pd). The residue was recrystallized after purification with GPC.

$[Pt_1Au_{37}(SC_2H_4Ph)_{24}]^0$ ($\mathbf{Pt_1Au_{22}}^0$).

A vial was charged with $Pt_1Au_{24}(PET)_{18}$ (9.2 mg, 1.25 μmol), followed by DCM (1.1 mL). Ethanolic solution containing 1 equation of $NaBH_4$ (0.1 M, 12.5 μL) was added to the vial (solution A′). One equation of $[Au_9(PPh_3)_8](NO_3)_3$ (5.0 mg, 1.25 μmol) was dissolved in DCM (275 μL). Ethanolic solution containing 1 equation of $NaBH_4$ (0.1 M, 12.5 μL) was rapidly added to the vial (solution B). Solution B was added dropwise to solution A′ and the mixture was stirred at room temperature for 20 min. Twelve equation of $AuClSMe_2$ (4.4 mg, 15 μmol) was put in THF (250 μL) in another vial to obtain a white suspension. Six equation of PET (1.01 μL, 7.5 μmol) was added dropwise to the suspension while stirring to obtain a pale yellow solution (solution C). Solution C was added dropwise to the mixture of solutions A′ and B and the mixture was stirred at room temperature for 130 min. Then, 12 equation of NEt_3 (2.08 μL, 15 μmol) was added to the solution and the mixture was stirred for 10 min. Finally, MeOH was added and the solution was evaporated to dryness. The residue was washed with EtOH and extracted with DCM. The extract was evaporated and then purified with Al_2O_3 column chromatography using the same procedure as that for $\mathbf{Pd_1Au_{22}}^0$. The isolated yield was 1.8 mg (13% with respect to Pt).

5.2.3 Photoelectron Spectroscopy

Anion PES was conducted by using a home-built apparatus equipped with an electrospray ionization (ESI) source, a time-of-flight mass spectrometer (TOF–MS), and a magnetic-bottle type photoelectron spectrometer (MB-PES) as described in Chap. 2. $\mathbf{M_1Au_{22}}^-$ were prepared by reducing 0.2 mg of $\mathbf{M_1Au_{22}}^0$ dissolved in THF (20 μL) with 1 equation of $NaBH_4$ dissolved in cold EtOH (0.01 M, 1.86 μL) and then diluted with THF (1 mL). $\mathbf{M_1Au_{22}}^-$ were introduced into the apparatus via the ESI source and desolvated by a heated capillary. Ions were transferred to a quadrupole linear ion trap (QLIT) through a skimmer (1 mm diameter). Continuously flowing ions were accumulated in the QLIT for 99 ms and released 300 μs before the acceleration of the TOF–MS. The released ions were accelerated to ~13 keV by a set of Wiley–McLaren type electrodes and were temporally and spatially separated by the TOF–MS with a mass resolution ($m/\Delta m$) of ~600. Pulsed third or fourth harmonics of an Nd:YAG nanosecond laser (Quanta-Ray GCR130, Spectra-Physics, US) or pulsed ArF excimer laser (PSX-100, MPB Communications Inc.) clipped by a pinhole were irradiated to the mass-selected ion beam during the TOF–MS. The emitted photoelectrons were collected by an inhomogeneous magnetic field generated by a permanent magnet and a coil. The kinetic energies of the photoelectrons were determined by the TOF of the electrons and then converted to the electron binding energies (EBEs). The EBEs were calibrated with the PE spectrum of I^- produced by the dissociative electron attachment to CH_3I. The energy resolution was approximately 110 mV for an electron with a kinetic energy of 1 eV. The obtained PE spectra were smoothed by averaging the data points within ±50 meV. The adiabatic electron affinity was determined by the energy of the spectral onset where the tangential line at the inflection point intersected with the horizontal axis. Laser fluences were measured at the exit of the Nd:YAG laser.

5.2.4 *Theoretical Calculation*

Density functional theory (DFT) calculation was conducted on $Au_{38}(SCH_3)_{24}{}^0$ (**Au_{23}(m)^0**), $[Pd_1Au_{37}(SCH_3)_{24}]^-$ (**Pd_1Au_{22}(m)$^-$**), and $[Pt_1Au_{37}(SCH_3)_{24}]^-$ (**Pt_1Au_{22}(m)$^-$**) with B3LYP functionals and 6–31G(d) basis sets for H, C, and S atoms and LanL2DZ basis sets for Pd, Pt, and Au atoms by using the Gaussian 16 program [33]. All structures were created from crystal structures by substituting the PET ligands with SCH_3 and structural optimization was performed. Frequency calculations on the optimized structures did not show any imaginary frequencies, indicating that the obtained structures were at local minima. Vertical detachment energies were calculated by subtracting the total energy of the structure-optimized **Pd_1Au_{22}(m)$^-$** and **Pt_1Au_{22}(m)$^-$** from that of the neutrals with the fixed geometries without considering the zero-point energies. The contributions of atomic orbitals to each Kohn–Sham orbital were analyzed based on Mulliken population analysis (MPA) using the GaussSum program [34]. NBO charge analysis was conducted by the NBO program implemented in the Gaussian [35].

5.3 Results and Discussion

5.3.1 *Synthesis and Characterization*

The synthesis and characterization was conducted by Ms. Emi Ito. I will briefly summarize her results for a better understanding of anion PES and DFT calculation.

Neutral precursor clusters **$M_1Au_{22}{}^0$** (M = Pd, Pt) were synthesized by using the fusion reactions between the hydride-doped Au cluster $[HAu_9(PPh_3)_8]^{2+}$ (**Au_9–H^{2+}**) and the PET-protected clusters $[M_1Au_{24}(PET)_{18}]^-$ (**$M_1Au_{12}{}^-$**) [30–32]: two DCM solutions containing the equivalent amount of **Au_9–H^{2+}** and **$M_1Au_{12}{}^-$** (M = Pd, Pt) were mixed, followed by the addition of 6 equation of Au(I)-PET complex.

The composition of the product was confirmed by positive-mode matrix-assisted laser desorption/ionization (MALDI) mass spectrometry (Fig. 5.1a), EPR spectroscopy (Fig. 5.1b), UV–Vis absorption spectroscopy (Fig. 5.1c), and SCXRD (Fig. 5.1d). The MALDI mass spectrum of $\mathbf{Pd_1Au_{22}}^0$ (green trace in Fig. 5.1a) indicates the successful synthesis, while the MALDI mass spectrum of $\mathbf{Pt_1Au_{22}}^0$ (pink trace in Fig. 5.1a) could not distinguish $\mathbf{Pt_1Au_{22}}^0$ (calculated *m/z* 10,776.25) and the undoped reference $\mathbf{Au_{23}}^0$ (*m/z* 10,778.14) due to the limited resolution. The EPR spectrum of $\mathbf{Pt_1Au_{22}}^0$ recorded in a frozen DCM at 5 K showed that both $\mathbf{Pd_1Au_{22}}^0$ and $\mathbf{Pt_1Au_{22}}^0$ are paramagnetic and have an open electronic structure, whereas $\mathbf{Au_{23}}^0$ is diamagnetic (Fig. 5.1b). This EPR spectrum agreed well with that of $\mathbf{Pt_1Au_{22}}^0$ recently synthesized by a co-reduction method [36]. The optical spectra of $\mathbf{M_1Au_{22}}^0$ exhibited a similar profile to that of $\mathbf{Au_{23}}^0$, except for an additional band at the low energy region of <1 eV (Fig. 5.1c) due to the electronic transition to a singly occupied molecular orbital (SOMO), which also implied the open electronic shells. This result indicates that the electronic structures of $\mathbf{Au_{23}}^0$ and $\mathbf{M_1Au_{22}}^0$ are similar, except that the electronic shell is closed or open. Figure 5.1d represents the SCXRD results for $\mathbf{Pd_1Au_{22}}^0$ and $\mathbf{Pt_1Au_{22}}^0$ obtained at 103 and 100 K, respectively. The absence of counter ions supported that $\mathbf{Pd_1Au_{22}}^0$ and $\mathbf{Pt_1Au_{22}}^0$ are neutral in charge state and electronically open. The structures look nearly identical to that of $\mathbf{Au_{23}}^0$: they are composed of bi-icosahedral MAu_{22} (M = Pd, Pt) cores protected by three $Au(PET)_2$ and six $Au_2(PET)_3$ oligomers [3, 24]. The location of the Pd dopant in $\mathbf{Pd_1Au_{22}}^0$ was identified at one of the centers of two icosahedrons [24, 37], whereas that of the Pt dopant in $\mathbf{Pt_1Au_{22}}^0$ was not identified because Au and Pt could not be distinguished by SCXRD. Nevertheless, it is most likely that the Pt dopant occupied either center of the bi-icosahedrons based on the previous examples in which Pt exclusively took the central site of Au superatoms [24, 31, 32, 36, 38–43]. In summary, the structures of $\mathbf{M_1Au_{22}}^0$ (M = Pd, Pt) are nearly identical to that of $\mathbf{Au_{23}}^0$.

Finally, $\mathbf{M_1Au_{22}}^0$ (M = Pd, Pt) was reduced by the reaction with $NaBH_4$ [31, 44]. Negative-mode ESI mass spectra after the reduction exhibit a single peak (Fig. 5.2a). Analysis of the isotope patterns confirmed the assignment to $\mathbf{M_1Au_{22}}^-$. Since the structures of $\mathbf{M_1Au_{22}}^-$ could not be determined by SCXRD, the structural information was obtained by optical spectroscopy and DFT calculations. The UV–Vis spectra of $\mathbf{M_1Au_{22}}^-$ showed similar profiles with the corresponding $\mathbf{M_1Au_{22}}^0$ except that the lowest energy peak observed in $\mathbf{M_1Au_{22}}^0$ disappeared due to suppression of the optical transition to SOMO upon the reduction (Fig. 5.2b).

5.3.2 DFT Calculation

DFT calculation was conducted to compare the geometry and electronic structure between $\mathbf{MAu_{22}}^-$ and $\mathbf{Au_{23}}^0$, while the single crystal of $\mathbf{MAu_{22}}^-$ was not obtained yet. DFT calculation was conducted on model systems $[Pd_1Au_{37}(SCH_3)_{24}]^-$ ($\mathbf{Pd_1Au_{22}(m)^-}$), $[Pt_1Au_{37}(SCH_3)_{24}]^-$ ($\mathbf{Pt_1Au_{22}(m)^-}$), and $Au_{38}(SCH_3)_{24}$ ($\mathbf{Au_{23}(m)^0}$)

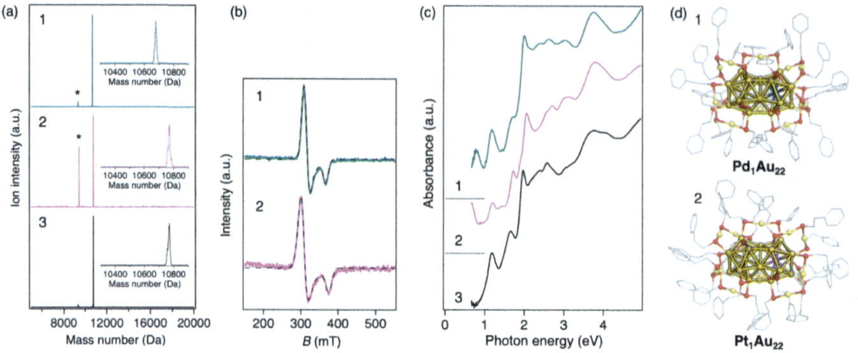

Fig. 5.1 **a** Positive-mode MALDI mass spectra of $\mathbf{Pd_1Au_{22}}^0$ (green), $\mathbf{Pt_1Au_{22}}^0$ (pink), and $\mathbf{Au_{23}}^0$ (black). Peaks with an asterisk are assigned to laser-induced fragments ($Au_4(PET)_4$ loss). The inset shows the expanded mass spectra. of $\mathbf{Pd_1Au_{22}}^0$ and $\mathbf{Pt_1Au_{22}}^0$. **b** Experimental (colored, solid) and simulated (black, dotted) X-band EPR of $\mathbf{Pd_1Au_{22}}^0$ (green) and $\mathbf{Pt_1Au_{22}}^0$ (pink). **c** Optical absorption spectra of the products of $\mathbf{Pd_1Au_{22}}^0$ (green), $\mathbf{Pt_1Au_{22}}^0$ (pink), and $\mathbf{Au_{23}}^0$ (black) in $CHCl_3$ as a function of the photon energy. **d** Crystal structures of $\mathbf{Pd_1Au_{22}}^0$ and $\mathbf{Pt_1Au_{22}}^0$. Metal and sulfur atoms are shown as spheres. Color code: yellow, Au; teal, Pd; pink, Pt; red, S; gray, C. Organic residues are depicted as gray sticks, and H atoms and disordered parts are omitted for clarity. The central position is treated as half occupancy of the Pd/Au and Pt/Au atoms, respectively

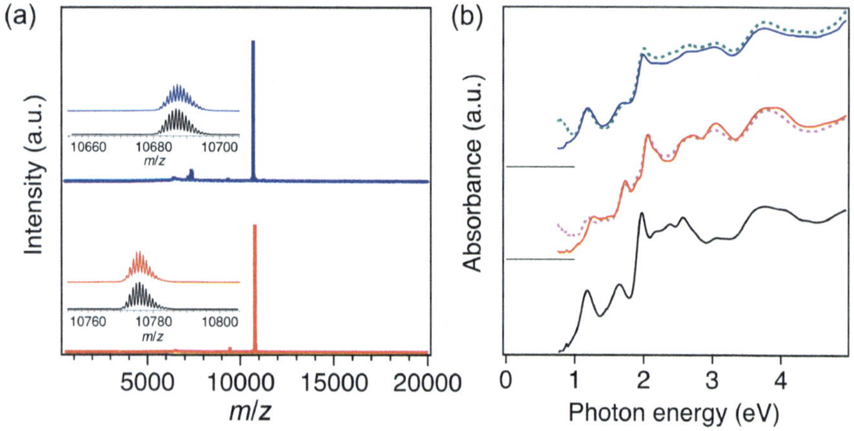

Fig. 5.2 **a** Negative-ion ESI mass spectra of $\mathbf{Pd_1Au_{22}}^-$ (blue) and $\mathbf{Pt_1Au_{22}}^-$ (red). Insets compare the experimental (colored) and simulated (black) isotope patterns. **b** Comparison of UV–Vis spectra of $\mathbf{Pd_1Au_{22}}^-$ (blue, solid) and $\mathbf{Pt_1Au_{22}}^-$ (red, solid) to that of $\mathbf{Au_{23}}^0$ (black, solid) in THF. Dotted lines show the spectra of $\mathbf{Pd_1Au_{22}}^0$ (green) and $\mathbf{Pt_1Au_{22}}^0$ (pink)

by simplifying PET to methane thiolate to reduce the calculation cost. The DFT-optimized structures are presented in Fig. 5.3. The DFT-optimized structure of $\mathbf{Au_{23}(m)}^0$ reproduced the SCXRD structure of $\mathbf{Au_{23}}^0$, demonstrating the reliability of the theoretical method used here. The optimized structures of $\mathbf{M_1Au_{22}(m)}^-$ were similar to the crystal structures of corresponding neutrals $\mathbf{M_1Au_{22}}^0$ (Fig. 5.1d). These

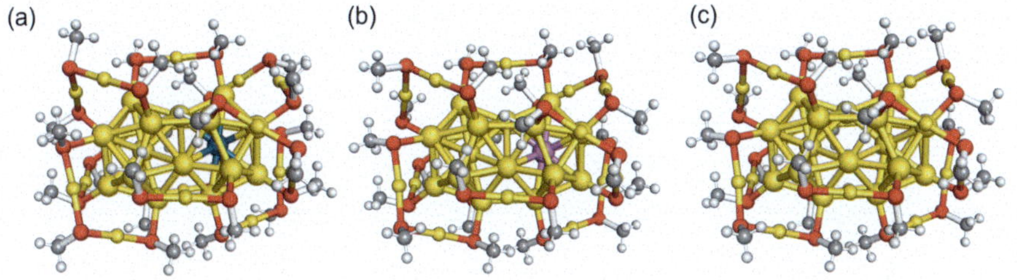

Fig. 5.3 Optimized structures of **a** $\mathbf{Pd_1Au_{22}(m)^-}$, **b** $\mathbf{Pt_1Au_{22}(m)^-}$, and **c** $\mathbf{Au_{23}(m)^0}$

results indicate that $\mathbf{M_1Au_{22}}^-$ retains the geometric structures after the reduction and isostructural to $\mathbf{Au_{23}}^0$.

The similarity of UV–visible absorption spectra between $\mathbf{M_1Au_{22}}^-$ and $\mathbf{Au_{23}}^0$ (Fig. 5.2b) is rationalized by the similar energy diagrams of their Kohn–Sham orbitals, although the Kohn–Sham orbitals of $\mathbf{M_1Au_{22}}^-$ are upshifted with respect to the corresponding orbitals of $\mathbf{Au_{23}}^0$ due to the negative charge (Fig. 5.4). The shapes of Kohn–Sham orbitals of LUMO + 2 to HOMO–4 of $\mathbf{M_1Au_{22}(m)}^-$ and $\mathbf{Au_{23}(m)}^0$ are shown in Fig. 5.5. It appears that the orbitals can be divided into five types of supervalence bonding: 2Σ*, 1Δ, 1Π*, 2Σ, and 1Π based on its shape (Fig. 5.6a), and the occupied orbitals are energetically divided into two subsets (Fig. 5.6b): (1) two 1Π* orbitals and (2) one 2Σ and two 1Π orbitals. The contribution of atomic orbitals to each KS orbitals of $\mathbf{M_1Au_{22}(m)}^-$ was analyzed and the projected density of states thus obtained (Fig. 5.7) indicates that the frontier KS orbitals are mainly composed of sp orbitals of the metals in the MAu_{22} core. The nature of the occupied orbitals of $\mathbf{M_1Au_{22}(m)}^-$ agrees with the theoretical prediction for $\mathbf{Au_{23}(m)}^0$ based on the SVB model [16, 17]. The NBO charge analysis of $\mathbf{M_1Au_{22}(m)}^-$ and $\mathbf{Au_{23}(m)}^0$ revealed that the charge distribution is not uniform in the hetero-bi-icosahedral M_1Au_{22} core. As shown in Table 5.2, the charge distribution at the MAu_{12} side of the core of $\mathbf{M_1Au_{22}(m)}^-$ was negatively biased by ~0.3*e* compared to the Au_{13} side because of the lower valency of Pd and Pt than Au. Nevertheless, we can conclude that $\mathbf{M_1Au_{22}}^-$ (M = Pd, Pt) and $\mathbf{Au_{23}}^0$ are isoelectronic and the electron configurations of both expressed as electronically homonuclear superatomic molecules: $(1\Sigma)^2(1\Sigma^*)^2(1\Pi)^4(2\Sigma)^2(1\Pi^*)^4$.

5.3.3 Photoelectron Spectroscopy

The PE spectra of the mass-selected $\mathbf{M_1Au_{22}}^-$ (M = Pd, Pt) were recorded to investigate the supervalence bonding. The representative PE spectra of $\mathbf{M_1Au_{22}}^-$ (M = Pd, Pt) are shown in Fig. 5.8a. The PE spectra of both clusters at 193 nm (1 mJ cm^{-2}) exhibited two peaks I and II centered at ~3.8 and ~4.3 eV along with intense broad bands at the higher EBE of >4.5 eV. The PE spectra at 266 nm showed peak I and a rising edge of peak II (Fig. 5.8a). Based on the onset energies of peak I, the

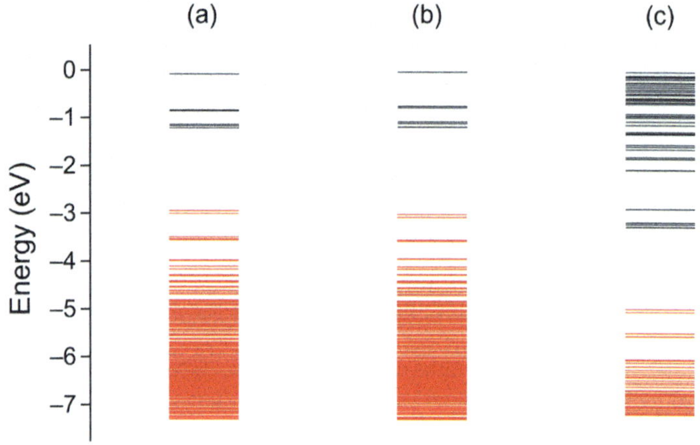

Fig. 5.4 Energy diagram of Kohn–Sham orbitals of **a** $\mathbf{Pd_1Au_{22}(m)^-}$, **b** $\mathbf{Pt_1Au_{22}(m)^-}$, and **c** $\mathbf{Au_{23}(m)^0}$. Red and black lines indicate occupied orbitals and unoccupied orbitals, respectively

adiabatic electron affinities of corresponding neutrals $\mathbf{M_1Au_{22}}^0$ were determined to be 3.53 ± 0.06 and 3.60 ± 0.06 eV for M = Pd and Pt, respectively. DFT calculation of $\mathbf{M_1Au_{22}(m)^-}$ (M = Pd, Pt) showed that the vertical detachment energies of $\mathbf{M_1Au_{22}(m)^-}$ were 3.64 and 3.72 eV for M = Pd and Pt respectively and qualitatively reproduced the larger electron affinity of $\mathbf{Pt_1Au_{22}}^0$ than $\mathbf{Pd_1Au_{22}}^0$. The density of states of $\mathbf{M_1Au_{22}(m)^-}$ (M = Pd, Pt) was obtained by assuming the identical photoelectron detachment probability from all the states exhibited two humps at the onset region due to the supervalence bonding orbitals and the intense broadband at the high EBE (Fig. 5.8b). The density of states reproduced the experimental spectral features shown in Fig. 5.8a and indicated that peak I was assigned to the PE detachment from nearly degenerate 1Π* orbitals, whereas peak II was assigned to that from nearly degenerate 2Σ and 1Π orbitals. These PE results supported the SVB theory.

5.4 Summary

In summary, a super valence bond model in homonuclear superatomic molecules proposed for a bi-icosahedral Au_{23} core in $Au_{38}(PET)_{24}$ was tested by gas-phase photoelectron spectroscopy of $[M_1Au_{37}(PET)_{24}]^-$ (M = Pd, Pt) having similar electronic structures. Targeted anions were obtained by the reduction of neutral $[M_1Au_{37}(PET)_{24}]^0$ synthesized by Ms. Emi Ito via fusion reactions between two superatoms, $[HAu_9(PPh_3)_8]^{2+}$ and $[M_1Au_{24}(PET)_{18}]^-$. The anion photoelectron spectra of $[M_1Au_{37}(PET)_{24}]^-$ showed two peaks with the electron binding energies of ~3.8 and ~4.5 eV, which were assigned to the PE detachment from the 1Π* and 1Π/2Σ orbitals. This study not only demonstrates the wide applicability of the fusion reactions of superatoms for the targeted synthesis of superatomic molecules, but also

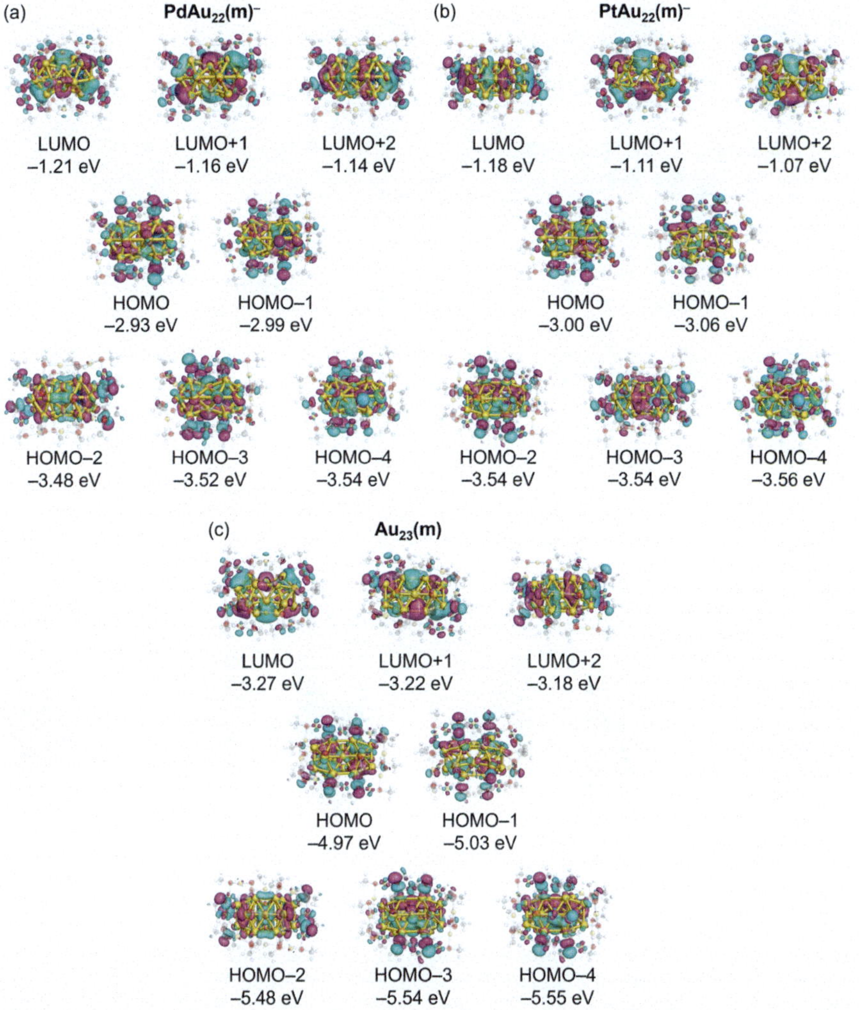

Fig. 5.5 The shape of Kohn–Sham orbitals of **a** $\mathbf{Pd_1Au_{22}(m)^-}$, **b** $\mathbf{Pt_1Au_{22}(m)^-}$, and **c** $\mathbf{Au_{23}(m)^0}$. Isodensity value was set to 0.015

provides experimental support for the super valence bond model in homonuclear superatomic molecules.

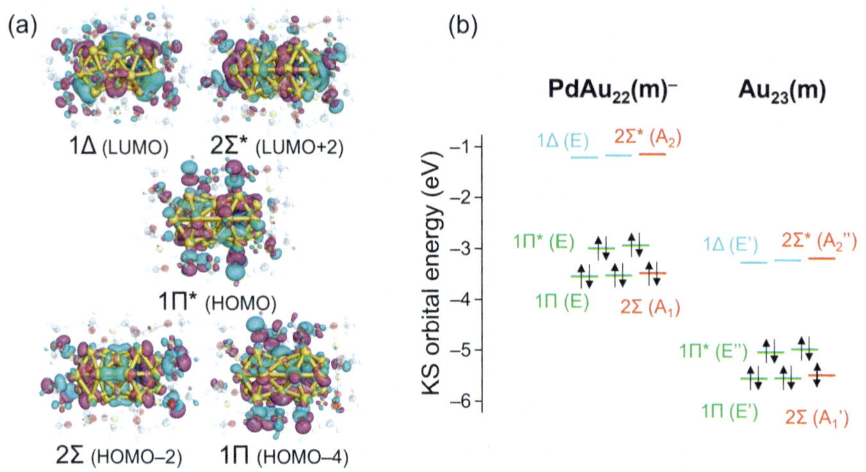

Fig. 5.6 **a** The shapes of selected Kohn–Sham orbitals of $\mathbf{Pd_1Au_{22}(m)^-}$. Isodensity value was set to 0.015. **b** Energy diagram of the selected KS orbitals of $\mathbf{Pd_1Au_{22}(m)^-}$ (left) and $\mathbf{Au_{23}(m)^0}$ (right)

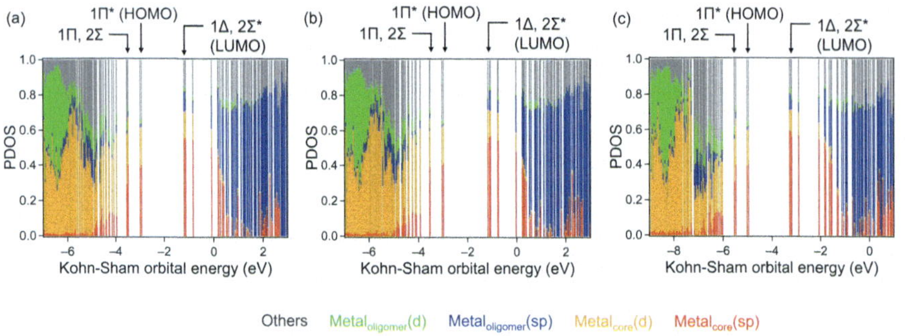

Fig. 5.7 Projected density of states of **a** $\mathbf{Pd_1Au_{22}(m)^-}$, **b** $\mathbf{Pt_1Au_{22}(m)^-}$, and **c** $\mathbf{Au_{23}(m)^0}$. Color code: black, atomic orbitals of S, C, and H; green, d orbitals of Au in oligomers; blue, sp orbitals of Au in oligomers; yellow, d orbitals of Au, Pd, and Pt in the core; red, sp orbitals of Au, Pd, and Pt in the core

Table 5.2 The result of NBO charge analysis

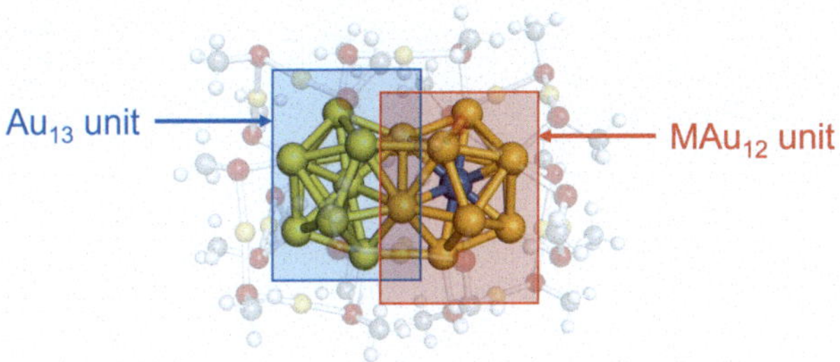

NBO charge	$Pd_1Au_{22}(m)^-$	$Pt_1Au_{22}(m)^-$	$Au_{23}(m)^0$
Au_{13} unit	–0.448	–0.471	–0.510
MAu_{12} unit	–0.771	–0.830	–0.511
Difference (MAu_{12}–Au_{13})	–0.323	–0.359	–0.001

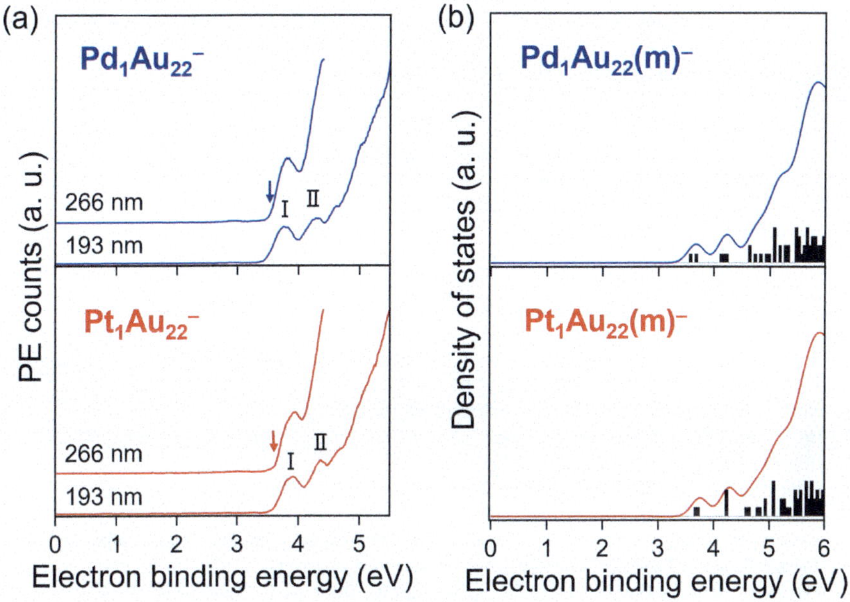

Fig. 5.8 **a** Gas-phase PE spectra of $\mathbf{Pd_1Au_{22}}^-$ (top) and $\mathbf{Pt_1Au_{22}}^-$ (bottom). Blue and red traces at 266 nm were measured with a laser fluence of 0.5 and 1 mJ cm^{-2}, respectively. Arrows show the adiabatic electron affinities. **b** Calculated density of states (line) and histograms of each energy level with a bin width of 50 meV (bars) of $\mathbf{Pd_1Au_{22}(m)}^-$ (top) and $\mathbf{Pt_1Au_{22}(m)}^-$ (bottom). Each energy level was broadened with full-width half-maximum of 0.35 eV

References

1. Teo BK, Zhang H (1995) Coord Chem Rev 143:611
2. Shichibu Y, Negishi Y, Watanabe T, Chaki NK, Kawaguchi H, Tsukuda T (2007) J Phys Chem C 111(111):7845
3. Qian H, Eckenhoff WT, Zhu Y, Pintauer T, Jin R (2010) J Am Chem Soc 132:8280
4. Wan X-K, Lin Z-W, Wang Q-M (2012) J Am Chem Soc 134:14750
5. Jin R, Liu C, Zhao S, Das A, Xing H, Gayathri C, Xing Y, Rosi NL, Gil RR, Jin R (2015) ACS Nano 9:8530
6. Takano S, Yamazoe S, Koyasu K, Tsukuda T (2015) J Am Chem Soc 137:7027
7. Song Y, Fu F, Zhang J, Chai J, Kang X, Li P, Li S, Zhou H, Zhu M (2015) Angew Chem Int Ed 54:8430
8. Jin S, Du W, Wang S, Kang X, Chen M, Hu D, Chen S, Zou X, Sun G, Zhu M (2017) Inorg Chem 56:11151
9. Shen H, Deng G, Kaappa S, Tan T, Han Y-Z, Malola S, Lin S-C, Teo BK, Häkkinen H, Zheng N (2019) Angew Chem Int Ed 58:17731
10. Yuan S-F, Li J-J, Guan Z-J, Lei Z, Wang Q-M (2020) Chem Commun 56:7037
11. Dong J, Gao Z-H, Zhang Q-F, Wang L-S (2021) Angew Chem Int Ed 60:2424
12. Kulkarni VK, Khiarak BN, Takano S, Malola S, Albright EL, Levchenko TI, Aloisio MD, Dinh C-T, Tsukuda T, Häkkinen H, Crudden CM (2022) J Am Chem Soc 144:9000
13. Shigeta T, Takano S, Tsukuda T (2022) Angew Chem Int Ed 61:e202113275
14. Li Y, Song Y, Zhang X, Liu T, Xu T, Wang H, Jiang D-E, Jin R (2022) J Am Chem Soc 144:12381
15. Nobusada K, Iwasa T (2007) J Phys Chem C 111:14279
16. Cheng L, Ren C, Zhang X, Yang J (2013) Nanoscale 5:1475
17. Cheng L, Yang J (2013) J Chem Phys 138:141101
18. Nishigaki J, Koyasu K, Tsukuda T (2014) Chem Rec 14:897
19. Tlahuice-Flores A, Munos-Castro A (2019) Int J Quantum Chem 119:e25756
20. Munos-Castro A (2019) Chem Commun 55:7307
21. Omoda T, Takano S, Tsukuda T (2021) Small 17:2001439
22. Xu C, Zhou Y, Yi J, Li D, Shi L, Cheng L (2022) J Phys Chem Lett 13:1931
23. Mingos DMP (2015) Dalton Trans 44:6680
24. Ito E, Takano S, Nakamura T, Tsukuda T (2021) Angew Chem Int Ed 60:645
25. Hirata K, Tomihara R, Koyasu K, Tsukuda T (2019) Phys Chem Chem Phys 21:17463
26. Veenstra AP, Monzel L, Baksi A, Czekner J, Lebedkin S, Schneider EK, Pradeep T, Unterreiner A-N, Kappes MM (2020) J Phys Chem Lett 11:2675
27. Koyasu K, Tsukuda T (2021) J Chem Phys 154:140901
28. Ito S, Tasaka Y, Nakamura K, Fujiwara Y, Hirata K, Koyasu K, Tsukuda T (2022) J Phys Chem Lett 13:5049
29. Brandys M-C, Jennings MC, Puddephatt RJ (2000) J Chem Soc Dalton Trans 4601
30. Takano S, Hirai H, Muramatsu S, Tsukuda T (2018) J Am Chem Soc 140:8380
31. Suyama M, Takano S, Nakamura T, Tsukuda T (2019) J Am Chem Soc 141:14048
32. Takano S, Ito S, Tsukuda T (2019) J Am Chem Soc 141:15994
33. Frisch MJ, Trucks GW, Schlegel HB, Scuseria GE, Robb MA, Cheeseman JR, Scalmani G, Barone V, Petersson GA, Nakatsuji H, Li X, Caricato M, Marenich AV, Bloino J, Janesko BG, Gomperts R, Mennucci B, Hratchian HP, Ortiz JV, Izmaylov AF, Sonnenberg JL, Young WD, Ding F, Lipparini F, Egidi F, Goings J, Peng B, Petrone A, Henderson T, Ranasinghe D, Zakrzewski VG, Gao J, Rega N, Zheng G, Liang W, Hada M, Ehara M, Toyota K, Fukuda R, Hasegawa J, Ishida M, Nakajima T, Honda Y, Kitao O, Nakai H, Vreven T, Throssell K, Montgomery Jr. JA, Peralta JE, Ogliaro F, Bearpark MJ, Heyd JJ, Brothers EN, Kudin KN, Staroverov VN, Keith TA, Kobayashi R, Normand J, Raghavachari K, Rendell AP, Burant JC, Iyengar SS, Tomasi J, Cossi M, Millam JM, Klene M, Adamo C, Cammi R, Ochterski JW, Martin RL, Morokuma K, Farkas O, Foresman JB, Fox DJ (2016) Gaussian 16, Revision B.01. Gaussian, Inc., Wallingford CT

34. O'Boyle NM, Tenderholt AL, Langner KM (2008) J Comput Chem 29:839
35. Glendening ED, Reed AE, Carpenter JE, Weinhold F, NBO, Ver. 3.1
36. Liu X, Wang E, Zhou M, Wan Y, Zhang Y, Liu H, Zhao Y, Li J, Gao Y, Zhu Y (2022) Angew Chem Int Ed 61:e202207685
37. Negishi Y, Igarashi K, Munakata K, Ohgake W, Nobusada K (2012) Chem Commun 48:660
38. Qian H, Jiang D-E, Li G, Gayathri C, Das A, Gil RR, Jin R (2012) J Am Chem Soc 134:16159
39. Kwak K, Yang Q, Kim M, Jiang D-E, Lee D (2015) J Am Chem Soc 137:10833
40. Tian S, Liao L, Yuan J, Yao C, Chen J, Yang J, Wu J (2016) Chem Commun 52:9873
41. Kim M, Tang Q, Kumar AVN, Kwak K, Choi W, Jiang D-E, Lee D (2018) J Phys Chem Lett 9:982
42. Fei W, Antonello S, Dainese T, Dolmella A, Lahtinen M, Rissanen K, Venzo A, Maran F (2019) J Am Chem Soc 141:16033
43. Hirai H, Takano S, Nakashima T, Iwasa T, Taketsugu T, Tsukuda T (2020) Angew Chem Int Ed 61:e202207290
44. Negishi Y, Chaki NK, Shichibu Y, Whetten RL, Tsukuda T (2007) J Am Chem Soc 129:11322

Chapter 6
Bonding Interaction in Heteronuclear Superatomic Molecules

Abstract The sixth chapter describes the bonding interaction in $[PdAu_{24}(C{\equiv}CAr^F)_{18-2n}]^{2-}$ (Ar^F = 3,5-$(CF_3)_2C_6H_3$) generated by the collision-induced dissociation (CID) of $[PdAu_{24}(C{\equiv}CAr^F)_{18}]^{2-}$ in the gas phase. The number of valence electrons of $[PdAu_{24}(C{\equiv}CAr^F)_{18-2n}]^{2-}$ was equal to $8 + 2n$ and did not agree with the discrete magic number of the spherical jellium model 8, 18, 20, and so on, suggesting the presence of the special binding mechanism of valence electrons. To elucidate the binding mechanism of excess valence electrons in CID products, anion photoelectron spectroscopy (PES) and density functional theory (DFT) calculation were conducted. Anion PES showed the gradual decrease of the electron binding energy with *n*. DFT calculation of the simplest fragment $[PdAu_{24}(C{\equiv}CAr^F)_{16}]^{2-}$ with 10 electrons showed that the removal of two ligands coordinating to the same Au atoms yielded the most plausible structure. The binding mechanism of the valence electrons in $[PdAu_{24}(C{\equiv}CAr^F)_{16}]^{2-}$ can be interpreted as the supervalence bonding of the $PdAu_{12}$ core with 8 electrons and the Au_2 unit with 2 electrons rather than a single superatom with 10 electrons. Therefore, $[PdAu_{24}(C{\equiv}CAr^F)_{16}]^{2-}$ was categorized to a heteronuclear superatomic molecule composed of superatoms with different electrons. The heteronuclear electron configuration persisted at the model structure of $[PdAu_{24}(C{\equiv}CAr^F)_6]^{2-}$ with 20 electrons.

Keywords Collision-induced dissociation · Reductive elimination · Superatomic molecule · Gold cluster

6.1 Introduction

Collision-induced dissociation (CID) coupled with mass spectrometry (MS) is a powerful analytical tool in chemistry and has been applied to a variety of targets including small molecules, peptides [1], natural products [2], and molecular and atomic clusters [3]. CID MS not only provides information on the structures and energetics of the targets but also can be used for the gas-phase synthesis of novel

S. Ito, *Electronic Structures and Bonding Interaction of Structurally Defined Gold/Silver Superatoms*, Springer Theses, https://doi.org/10.1007/978-981-96-2892-6_6

metastable species [4] and the promotion of chemical reactions on metal complexes or clusters [5]. Recently, the targets of CID MS have been expanded to chemically synthesized ligand-protected metal clusters [6–16]. CID MS of protected metal clusters has yielded a molecular-level understanding of the bonding nature of the ligands and stabilities with the help of the geometrical structures determined by single-crystal X-ray crystallography [17, 18] and theoretical calculations [19, 20]. For example, CID MS revealed the relative bond strengths of different ligands on Au clusters [12, 15] and the formation of size-selected bare Ag clusters from hydride-rich Ag clusters [9, 11, 14]. The stabilities of the precursor Au clusters [17, 18, 21–24] and the product ions of CID [6, 12, 15] have been successfully explained by the superatom model [25]. High stability is attained when the superatomic orbitals are closed with specific numbers of valence electrons (n^*): the n^* values required to close the electronic shells for spherical Au clusters are 2, 8, 18, 20, 34, and so on [26, 27]. For example, eight valence electrons (8e) in an icosahedral $Au@Au_{12}$ superatomic cores protected by thiolate [17, 18], diphosphine [21], alkynyl [22], and N-heterocyclic carbine [23, 24] take the electron configuration of $(1S)^2(1P)^6$. Previous studies showed that the stability of the fragments of thiolate- or phosphine-protected structurally defined superatoms was also governed by the electronic shell closure [6, 12, 15].

In my master course study, I found that the CID process of alkynyl-protected Au clusters, which can take a wide variety of interfacial structures by the combination of σ- and π-bonding modes of alkynyl to Au [28–31], is different from thiolate- or phosphine-protected ones. The CID of eight-electron superatoms $[MAu_{24}(C{\equiv}CAr^F)_{18}]^{2-}$ ($M = Pd, Pt$; $Ar^F = 3,5\text{-}(CF_3)_2C_6H_3$; referred to as $\mathbf{MAu_{12}-0^{2-}}$) [32] proceeded via a successive desorption of 1,3-diyne, a coupling product of two alkynyl ligands, as shown in the following chemical formula

$$\begin{aligned}\left[MAu_{24}\left(C\equiv CAr^F\right)_{18}\right]^{2-} &\rightarrow \left[MAu_{24}\left(C\equiv CAr^F\right)_{16}\right]^{2-} + Ar^FC\equiv C{-}C\equiv CAr^F\\ &\rightarrow \left[MAu_{24}\left(C\equiv CAr^F\right)_{14}\right]^{2-} + 2\times Ar^FC\equiv C{-}C\equiv CAr^F\\ &\rightarrow \ldots\end{aligned}$$

and yielded $[MAu_{24}(C{\equiv}CAr^F)_{18-2n}]^{2-}$ ($n = 1$–6; referred to as $\mathbf{MAu_{12}{-}2\mathit{n}^{2-}}$) (Fig. 6.1) [33]. The CID process was analogous to the reductive elimination in organometal complexes and the number of valence electrons in $\mathbf{MAu_{12}{-}2\mathit{n}^{2-}}$ was increased to $8 + 2n$ from 8, which cannot be explained by the magic number of spherical jellium model 2, 8, 18, 20, and so on. One of the possible explanations for the unusual number of electrons is that the icosahedral MAu_{12} core accepts the increased electrons into 1D orbitals and grows in size by incorporating liberated Au atoms initially coordinated by alkynyl ligands. Another explanation is that the MAu_{12} core retains the closed-shell electron configuration with 8 electrons while liberated Au atoms accommodate the increased electrons as "adatoms" on the core. Density functional theory (DFT) calculation on the model structure of products by using $C{\equiv}CCF_3$ ligands supported the later explanation: the products can be viewed as clusters of superatoms composed from $M@Au_{12}$(8e) core and n units of Au_2(2e).

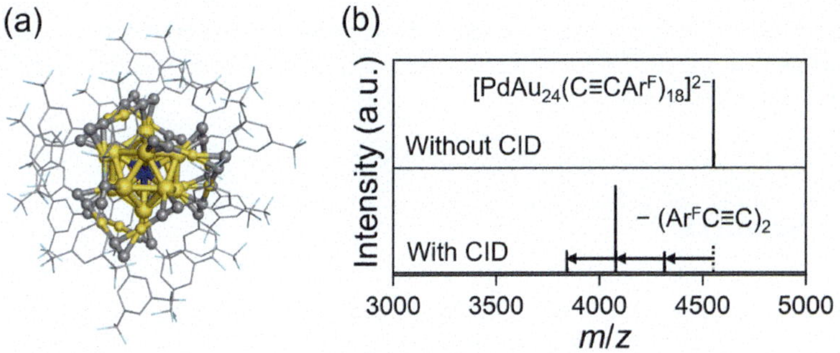

Fig. 6.1 **a** Crystal structure and **b** CID mass spectra of $[PdAu_{24}(C{\equiv}CAr^F)_{18}]^{2-}$

However, the experimental characterization of the CID fragment was not conducted because the ion trap was not installed yet in home-built apparatus.

The aim of this study is to elucidate the binding mechanism of unusual excess electrons in $[PdAu_{24}(C{\equiv}CAr^F)_{18-2n}]^{2-}$ (**PdAu$_{12}$–2*n*$^{2-}$**((8 + 2*n*)e)) generated by the CID of $[PdAu_{24}(C{\equiv}CAr^F)_{18}]^{2-}$ (**PdAu$_{12}$–0^{2-}**(8e)). To this end, I focused on the structural survey of the simplest fragment **PdAu$_{12}$–2^{2-}**(10e) by using anion photoelectron spectroscopy (PES) and DFT calculation. Anion PES revealed that the electron binding energy of **PdAu$_{12}$–2^{2-}**(10e) was decreased from that of **PdAu$_{12}$–0^{2-}**(8e). The comparison of PES and DFT calculation including the complete structure of ligands allowed us to conclude that the simplest product **PdAu$_{12}$–2^{2-}**(10e) can be viewed as a heteronuclear superatomic molecule: the cluster was composed from the icosahedral MAu_{12} core with 8 electrons and the Au_2 unit with 2 electrons rather than a single superatom with 10 electrons. This electron configuration is categorized as heteronuclear superatomic molecules unprecedented in the wet chemical synthesis.

6.2 Methods

6.2.1 General

All solvents except for tetrahydrofuran, sodium borohydride ($NaBH_4$), and triethylamine were purchased from FUJIFILM Wako Pure Chemical Corporation. 3,5-Bis(trifluoromethyl)ethynylbenzene (3,5-$(CF_3)_2C_6H_3C{\equiv}CH$, $Ar^FC{\equiv}CH$) and tetraphenylphosphonium chloride (PPh_4Cl) were purchased from Tokyo Chemical Industry. Tetrahydrofuran (THF) and phenylacetylene ($PhC{\equiv}CH$) were purchased from Kanto Chemicals. $[Au_{25}(C{\equiv}CAr^F)_{18}](PPh_4)$, $[PdAu_8(PPh_3)_8](NO_3)_2$, and $(Me_2S)AuCl$ were synthesized according to the reported procedures [32, 34–36]. Matrix-assisted laser desorption ionization mass spectrometry (MALDI-MS) of

$[Au_{25}(C{\equiv}CAr^F)_{18}](PPh_4)$ was measured by a Shimadzu AXIMA-CFR mass spectrometer using trans-2-[3-(4-tert-butylphenyl)-2-methyl-2-propenylidene] malononitrile (DCTB) as a matrix.

6.2.2 *Synthesis*

Synthesis of $[PdAu_{24}(C{\equiv}CAr^F)_{18}](PPh_4)_2$ was conducted based on the previous report [32]. $(Me_2S)AuCl$ (35.8 mg, 120 μmol) was suspended in THF (2 mL) and stirred in the dark. $Ar^FC{\equiv}CH$ (15.3 μL, 90 mol) and triethylamine (12.6 μL, 90 μmol) were added, resulting in a colorless solution (solution **A**) containing Au-C≡CR polymer complexes. $[PdAu_8(PPh_3)_8](NO_3)_2$ (19.4 mg, 5 μmol) was dissolved in a mixed solution of EtOH (500 μL) and THF (5 mL) in another test tube and the solution was cooled in an ice bath. Then, an ethanolic solution of $NaBH_4$ (0.10 M, 50 μL) was added to this chilled solution and after 30 s, solution **A** was added dropwise over 2–3 min. After stirring for 30 min, PPh_4Cl (3.7 mg, 10 μmol) in MeOH (1 mL) was added to the solution and the stirring was continued for 10 min. The crude mixture was poured into cold hexane (160 mL) and stirred for 1 h. The precipitate was collected by centrifugation and washed by cold hexane for three times. The resulting crude product was extracted with an EtOH:toluene = 1:1 mixture (3 mL) and reprecipitated by cold hexane. The precipitate was collected by centrifugation and extracted with Et_2O (4 mL). The supernatant was precipitated by hexane and dried in vacuo. The yield was 20.7 mg (43% based on the initial Pd). The product was used for the characterization without further crystallization.

6.2.3 *Photoelectron Spectroscopy*

Anion PES was conducted by using a home-built apparatus equipped with an electrospray ionization (ESI) source, a time-of-flight mass spectrometer (TOF–MS), and a magnetic-bottle type photoelectron spectrometer (MB-PES) introduced in Chap. 2. $[PdAu_{24}(C{\equiv}CAr^F)_{18}](PPh_4)_2$ samples dispersed in the 1:1 mixture of toluene and acetonitrile (~0.1 mg/mL) were introduced into the apparatus via the ESI source (bias voltage = –3 kV, flow rate = 1.5 μL/min) and desolvated by a capillary resistively heated up to 100 °C. Collision-induced dissociation occurred just after entering the vacuum chamber (Fig. 6.2) at the pressure of ~200 Pa by applying the high negative voltage on a focusing ring and a desolvation capillary (up to 320 V). CID occurred between the focusing ring and the skimmer 1 and its voltage difference is referred as the CID voltage. After the CID region, products were stored in a quadrupole linear ion trap (QLIT) for ~98 ms and released 400 μs before the acceleration of TOF–MS. Released ions were accelerated up to ~2.3 keV by a set of Wiley–McLaren type electrodes and temporally and spatially separated by TOF–MS. For the PES measurement, pulsed fifth harmonics (213 nm, 5.82 eV)

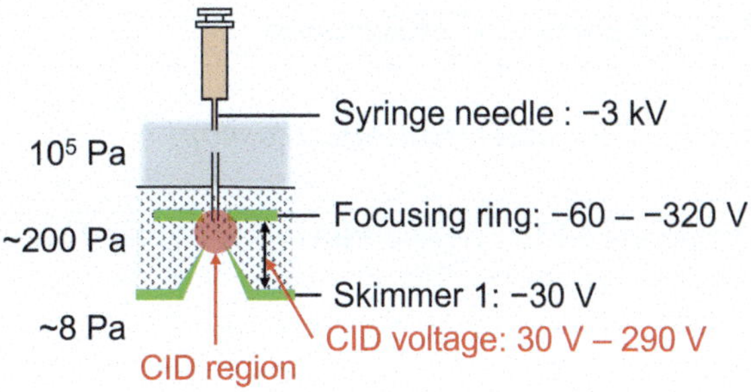

Fig. 6.2 Schematic illustration of the CID region of the PE spectrometer

of Nd:YAG laser (LS-2134UT, LOTIS TII) were irradiated to the mass-selected ion beam of $[PdAu_{24}(C{\equiv}CAr^F)_{18-2n}]^{2-}$ during the TOF–MS. Emitted photoelectrons were collected by an inhomogeneous magnetic field generated by a permanent magnet and a coil. The kinetic energies of photoelectrons were determined by TOF of electrons and then converted to the electron binding energies (EBE). Ions were switched on and off by each laser shot to subtract the background PE signal due to the PE emission from the surface of the chamber in a multichannel scaler. The calibration of EBE was done with the PE spectrum of I^- produced by the dissociative electron attachment to CH_3I. The PE spectra presented in this paper were obtained by accumulating ~30,000 laser shots including the background subtraction. The obtained PE spectra were smoothed by averaging neighboring data points in ±75 meV. The adiabatic electron affinity was determined by the energy of the spectral onset at which the second derivative of the PE spectrum took a local maximum. The standard deviation of several measurements was used as an error.

6.2.4 Theoretical Calculation

DFT calculation was conducted with PBE functionals, 6–31G(d) basis sets for C, H, and F, and LanL2DZ basis sets with a pseudopotential for Au and Pd by using the Gaussian 16 program [37]. Structural optimization was conducted with fixed spin multiplicity in the singlet state. The vibrational frequency calculation did not show the imaginary frequency, indicating that the optimized structures were located at the local minimum. The vertical detachment energies (VDEs) were calculated by subtracting the energy of the structure-optimized doubly charged anion from that of the singly charged anion with the fixed geometry without considering the zero-point energy. The contribution of each atomic orbital to the molecular orbitals was analyzed by using GaussSum program [38]. NBO charge analysis was conducted by the NBO program implemented in the Gaussian [39].

6.3 Results and Discussion

6.3.1 Photoelectron Spectroscopy

PE spectra of CID products **$PdAu_{12}$–0^{2-}**(8e)–**$PdAu_{12}$–12^{2-}**(20e) were successfully recorded (Fig. 6.3a). The PE spectrum of the precursor **$PdAu_{12}$–0^{2-}** exhibited a distinct band A at 3.6 eV with an onset energy of 3.27 ± 0.05 eV, which was assigned to superatomic 1P orbitals in Chap. 4 [40, 41]. PE signals above 4.7 eV are decreased due to the repulsive Coulomb barrier which is unique to anion PES of multiply charged anions as introduced in Sect. 1.3 [42]. The repulsive force between photoelectrons and remaining anions produces a barrier and slow photoelectrons cannot detach from multiply charged anions. The PE spectrum of **$PdAu_{12}$–2^{2-}** showed a broad band with a hump at 3.4 eV and its onset energy was decreased to 3.10 ± 0.04 eV. The peak position was ambiguous but seemed to be similar to **$PdAu_{12}$–0^{2-}**. PE spectra of **$PdAu_{12}$–4^{2-}**– **$PdAu_{12}$–12^{2-}** also exhibited a broad but distinct band similar to that of **$PdAu_{12}$–0^{2-}**. The spectral onset shown as arrows in Fig. 6.3a was monotonically shifted to the low-energy region with respect to *n*, the number of the desorption of 1.3-diynes, except for **$PdAu_{12}$–8^{2-}**(16e). The exceptional behavior suggests the presence of the special structure in **$PdAu_{12}$–8^{2-}**(16e). Another feature is that the width of band A showed an odd–even alternation: the band A of **$PdAu_{12}$–4^{2-}**(12e), **$PdAu_{12}$–8^{2-}**(16e), and **$PdAu_{12}$–12^{2-}**(20e) was narrow and the dip after the band A was prominent while that of **$PdAu_{12}$–2^{2-}**(10e), **$PdAu_{12}$–6^{2-}**(14e), and **$PdAu_{12}$–10^{2-}**(18e) is wide and unclear. The odd–even alternation suggests the degenerated electronic structure at the compositions with 12e, 16e, and 20e.

6.3.2 Structure of [$PdAu_{24}(C{\equiv}CAr^F)_{16}]^{2-}$ with 10 Electrons

In order to elucidate the geometrical and electronic structure of CID products, the structural survey of isomers was conducted by using DFT calculation mainly on the simplest product **$PdAu_{12}$–2^{2-}**(10e). The overall structure of ligands (3,5-$(CF_3)_2C_6H_3C{\equiv}C$) was incorporated in the DFT calculation. Structural optimization of the precursor **$PdAu_{12}$–0^{2-}**(8e) from the crystal structure was converged around the initial structure (referred to as **$PdAu_{12}$–$0a^{2-}$**, Fig. 6.4a) and the vertical detachment energy (VDE) of electrons were calculated to be 2.73 eV. The density of states (DOS) of **$PdAu_{12}$–$0a^{2-}$** obtained by broadening each energy level using a Gauss function with FWHM of 350 meV were quantitatively reproduced the experimental results (Fig. 6.4b). The difference in the binding energy may be raised from the poor calculation level, while the better calculation level will be too time consuming for the structural survey of CID products.

The structural candidates of **$PdAu_{12}$–2^{2-}**(10e) were limited to several species owing to the D_{3h} symmetry of **$PdAu_{12}$–0^{2-}**(8e) and the equivalency of six

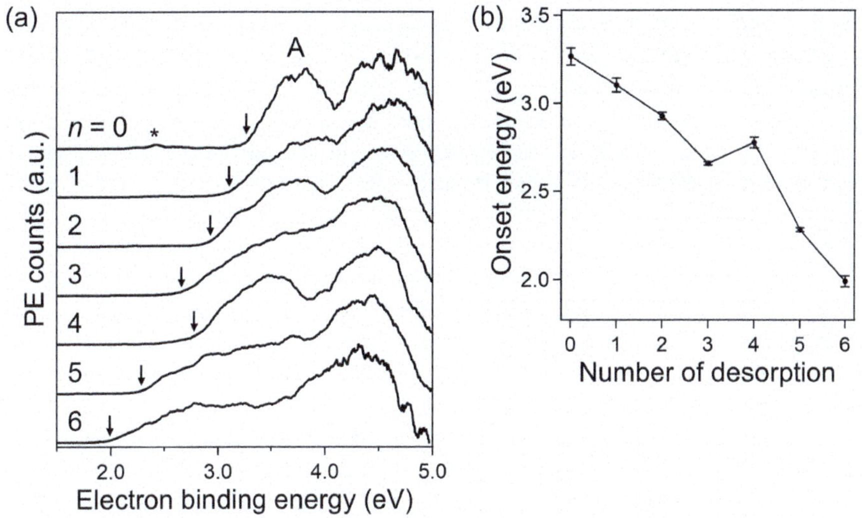

Fig. 6.3 **a** PE spectra and **b** the energy spectral onset of PE spectra of $\mathbf{PdAu_{12}{-}2}\boldsymbol{n}^{2-}$ (n = 0–6). Arrows indicate the position of the spectral onset. A peak with an asterisk is due to the two-photon PE detachment process via photoexcited states

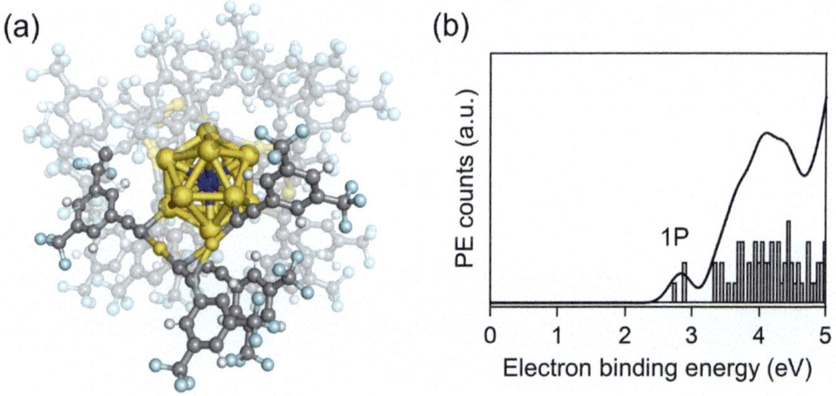

Fig. 6.4 **a** Optimized structure and **b** density of states of $\mathbf{PdAu_{12}{-}0a}^{2-}$. Bars in panel b show the histogram of energy levels with a bin width of 50 meV. Color code: yellow = Au; blue = Pd; gray = C; white = H; cyan = F

$Au_2(C{\equiv}CAr^F)_3$ oligomers. Initial structures were constructed by removing two alkynyl ligands from $\mathbf{PdAu_{12}{-}0a}^{2-}$ in a chemically intuitive manner as illustrated in Fig. 6.5. $\mathbf{PdAu_{12}{-}2a}^{2-}$, $\mathbf{PdAu_{12}{-}2b}^{2-}$, and $\mathbf{PdAu_{12}{-}2c}^{2-}$ were constructed by removing two ligands coordinating to the same Au atom, two ligands coordinating to the different Au atoms in a single $Au_2(C{\equiv}CAr^F)_3$ oligomer, and two ligands in different but neighboring two $Au_2(C{\equiv}CAr^F)_3$ oligomers, respectively. Relative energies (*ΔE*) and vertical detachment energies (VDEs) are listed in Table 6.1. It was found that **$PdAu_{12}$−2a^{2-}** was the most stable isomer and **$PdAu_{12}$−2b^{2-}** and **$PdAu_{12}$−2c^{2-}** was less stable by 0.34 and 0.44 eV, respectively. Regarding the geometrical structure, Au atoms initially coordinated by ligands (colored in green in Fig. 6.5) got close to each other to form a Au_2 unit and moved toward the MAu_{12} core. In case of **$PdAu_{12}$−2a^{2-}** and **$PdAu_{12}$−2b^{2-}**, the rest of five $Au_2(C{\equiv}CAr^F)_3$ oligomers and the icosahedral MAu_{12} core were retained, while the MAu_{12} core of **$PdAu_{12}$−2c^{2-}** was transformed to an anticuboctahedron from an icosahedron. Continuous shape measures (CSMs) [43, 44], which evaluate the distance between a set of points from ideal polyhedrons at CSM = 0, showed that the MAu_{12} core of **$PdAu_{12}$−2a^{2-}** and **$PdAu_{12}$−2b^{2-}** was a distorted icosahedron compared from **$PdAu_{12}$−0a^{2-}**, while **$PdAu_{12}$−2c^{2-}** was no longer an icosahedron but completely transformed to an anticuboctahedron (Table 6.1).

To assign the structure, the DOS of **$PdAu_{12}$–0a^{2-}** and **$PdAu_{12}$–2a^{2-}–$PdAu_{12}$–2c^{2-}** are compared in Fig. 6.6 in parallel with the comparison between PES of **$PdAu_{12}$–0^{2-}** and **$PdAu_{12}$–2^{2-}**. Features in the PE spectrum of **$PdAu_{12}$–2^{2-}**(10e) in comparison with **$PdAu_{12}$–0^{2-}**(8e) is that (1) the low-energy shift of the spectral onset from **$PdAu_{12}$–0^{2-}** (2) the similar peak position (3) a small hump between the spectral onset and the top of the peak. The DOS of **$PdAu_{12}$–2a^{2-}** fulfilled these features compared from that of **$PdAu_{12}$–0a^{2-}**, while the DOS of **$PdAu_{12}$−2b^{2-}** showed a distinct peak at the low EBE region and the DOS of **$PdAu_{12}$–2c^{2-}** did not shift to the low EBE region. This comparison between the experimental PE spectrum

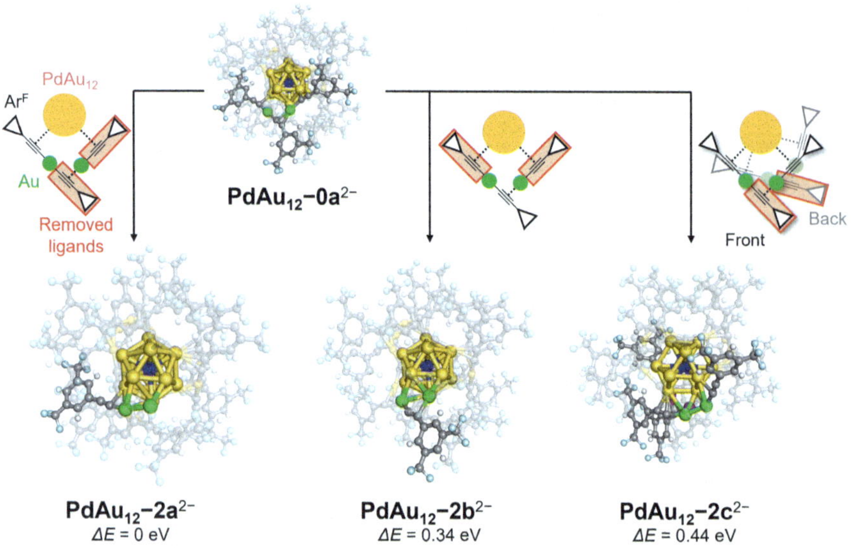

Fig. 6.5 Schematic illustration of the construction of initial structures and optimized structures of **$PdAu_{12}$−2a^{2-}**, **$PdAu_{12}$−2b^{2-}**, and **$PdAu_{12}$−2c^{2-}**. Green balls indicate Au atoms initially coordinated by removed ligands

Table 6.1 The relative energy (ΔE), the vertical detachment energy (VDE), and the continuous shape measure regarding icosahedrons (CSM_{Ih}) and anticuboctahedrons (CSM_{Acoc}) of **$PdAu_{12}$–$0a^{2-}$**, **$PdAu_{12}$–$2a^{2-}$**, **$PdAu_{12}$–$2b^{2-}$**, and **$PdAu_{12}$–$2c^{2-}$**

Structure	$PdAu_{12}$–$0a^{2-}$	$PdAu_{12}$–$2a^{2-}$	$PdAu_{12}$–$2b^{2-}$	$PdAu_{12}$–$2c^{2-}$
ΔE (eV)	0	0	0.34	0.44
VDE (eV)	2.72	2.44	2.22	2.74
CSM_{Ih}	0.057	0.273	0.335	5.247
CSM_{Acoc}	5.379	4.570	4.508	0.185

and the calculated DOS led us to conclude that the most stable isomer **$PdAu_{12}$–$2a^{2-}$** is the plausible structure of **$PdAu_{12}$–2^{2-}**. To support the structural assignment, ion mobility mass spectrometry (IM-MS), which evaluates the collision cross-section of ions in the gas phase, is undergoing as the international collaborative work.

To elucidate the binding mechanism of the excess two electrons in the most plausible structure **$PdAu_{12}$–$2a^{2-}$**, the electronic structure was analyzed. Kohn–Sham orbitals of **$PdAu_{12}$–$0a^{2-}$** and **$PdAu_{12}$–$2a^{2-}$** are shown in Fig. 6.7 by fixing the position of the Au_2 unit liberated from ligands and colored in green at the bottom. HOMO to HOMO–2 of **$PdAu_{12}$–$0a^{2-}$** were assigned to superatomic 1P orbitals distributed in the $PdAu_{12}$ core based on the shape of orbitals (Fig. 6.7a) and the high contribution (~30%) of s and p orbitals of the $PdAu_{12}$ core (Table 6.2). HOMO of

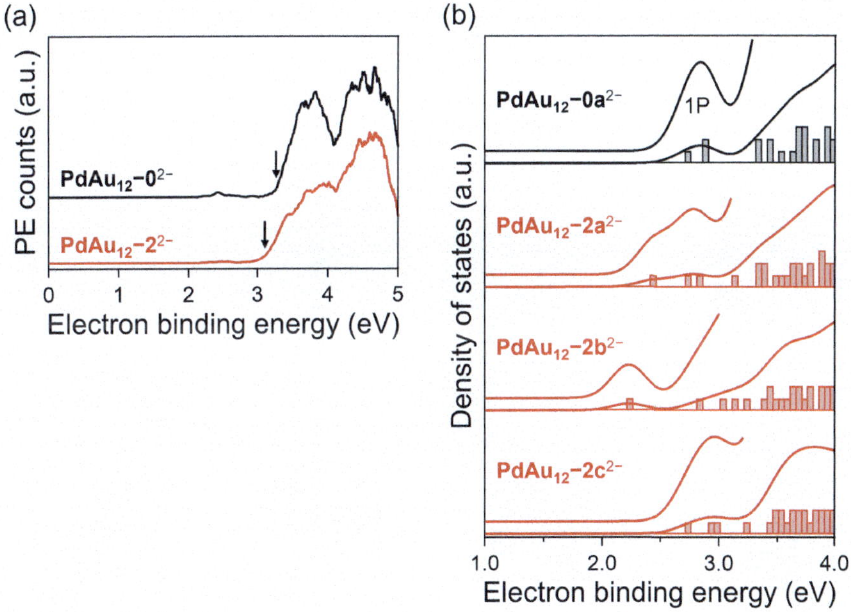

Fig. 6.6 **a** PE spectra of **$PdAu_{12}$–0^{2-}** and **$PdAu_{12}$–2^{2-}**. Arrows indicate the spectral onset. **b** Density of states of **$PdAu_{12}$–$0a^{2-}$**, **$PdAu_{12}$–$2a^{2-}$**, **$PdAu_{12}$–$2b^{2-}$**, and **$PdAu_{12}$–$2c^{2-}$**

PdAu$_{12}$–2a^{2-} (Fig. 6.7b) seemed to be delocalized over both the $PdAu_{12}$ core and the green Au_2 unit formed by the CID, while HOMO–1 and HOMO–2 were confined in the $PdAu_{12}$ core. Indeed, the contribution of sp orbitals of the Au_2 unit was significant in HOMO (24%) while it was negligible in HOMO–1 (2%) and HOMO–2 (1%) (Table 6.2). The biased contribution of the Au_2 unit suggested that the Au_2 units were electronically isolated from the $PdAu_{12}$ core. The coincidence of the energy level and the direction of electron distribution of HOMO–1 (−2.73 eV) and HOMO–2 (−2.84 eV) of **PdAu$_{12}$–2a^{2-}** and HOMO (−2.73 eV) and HOMO–1 (−2.88 eV) of **PdAu$_{12}$–0a^{2-}** (Fig. 6.7) also supported the non-interacting nature of HOMO–1 and HOMO–2 of **PdAu$_{12}$–2a^{2-}** with the Au_2 unit. Natural charge distribution analysis was consistent with the picture: a negative charge initially on two C≡CAr^F ligands (−0.87e) in **PdAu$_{12}$–0a^{2-}** was transferred mainly on the Au_2 unit in **PdAu$_{12}$–2a^{2-}** where the natural charge was decreased by 0.70e (Table 6.3). The Au–Au bond length between the Au_2 unit (2.69 Å) was shorter than that within the MAu_{12} core (2.98 ± 0.11 Å) and that between the Au_2 unit and $PdAu_{12}$ core (2.98 ± 0.07 Å) (Fig. 6.8), indicating the strong interaction in the Au_2 unit. Therefore, I concluded that the excess two electrons in **PdAu$_{12}$–2^{2-}**(10e) brought by the CID of **PdAu$_{12}$–0^{2-}**(8e) was accommodated by the Au_2 unit initially coordinated by desorbed ligands while the $PdAu_{12}$ core retained a closed-shell electron configuration with 8 electrons.

Electronic structure of **PdAu$_{12}$–2a^{2-}** could be rationalized by the super valence bonding (SVB) model [45, 46]. HOMO was assigned to σ* orbital made by the antibonding interaction between 1P orbital at the $PdAu_{12}$ core and 1S orbital at the Au_2 unit. HOMO–1 and HOMO–2 of **PdAu$_{12}$–2a^{2-}** were assigned to 1P orbitals with negligible interaction with the Au_2 unit. σ orbital between 1P and

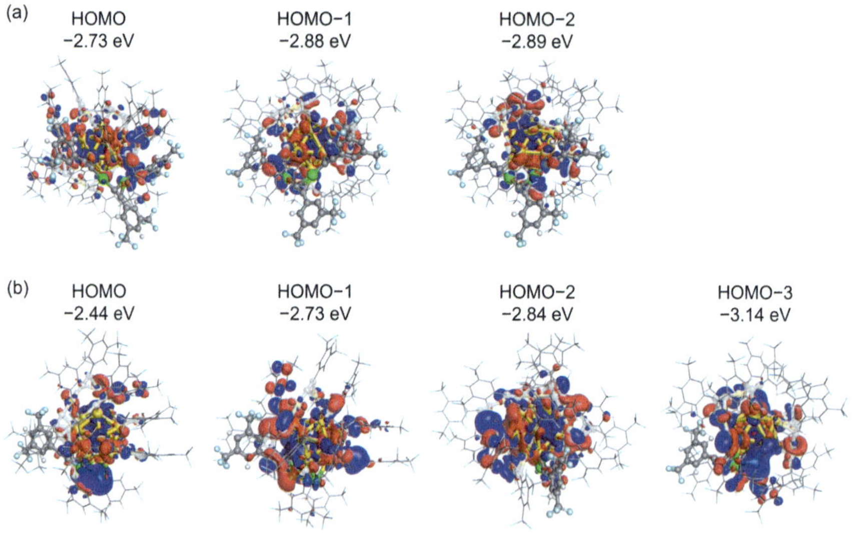

Fig. 6.7 Kohn–Sham orbitals of **a PdAu$_{12}$–0a^{2-}** and **b PdAu$_{12}$–2a^{2-}**. Contour values were set to 0.015

Table 6.2 Contribution analysis of atomic orbitals to molecular orbitals of $\mathbf{PdAu_{12}–0a^{2-}}$ and $\mathbf{PdAu_{12}–2a^{2-}}$

Molecule	Orbital	$PdAu_{12}$ core (sp)	Au_2 unit (sp)	Others
$\mathbf{PdAu_{12}–0a^{2-}}$	HOMO	29	–	71
	HOMO–1	31	–	69
	HOMO–2	31	–	69
$\mathbf{PdAu_{12}–2a^{2-}}$	HOMO	17	24	58
	HOMO–1	28	2	70
	HOMO–2	28	1	71
	HOMO–3	4	9	87

Table 6.3 Natural charge distribution analysis of $\mathbf{PdAu_{12}–0a^{2-}}$ and $\mathbf{PdAu_{12}–2a^{2-}}$

	$PdAu_{12}$ core	Two Au atoms coordinated by desorbed $C{\equiv}CAr^F$	$2 \times C{\equiv}CAr^F$	Others
$\mathbf{PdAu_{12}–0a^{2-}}$	−0.51	0.93	−0.84	−1.58
$\mathbf{PdAu_{12}–2a^{2-}}$	−0.67	0.20	–	−1.52
Difference ($\mathbf{PdAu_{12}–2a^{2-}}$–$\mathbf{PdAu_{12}–0a^{2-}}$)	−0.16	−0.73	0.84	0.06

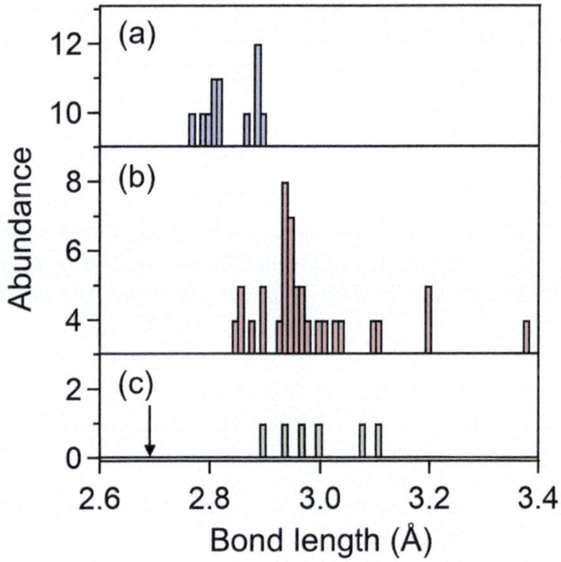

Fig. 6.8 A histogram of the bond length between **a** Pd–Au_{shell}, **b** Au_{shell}–Au_{shell}, and **c** Au_{shell}–Au_{Au2} in $\mathbf{PdAu_{12}–2a^{2-}}$. An arrow indicates the bond length in the Au_2 unit

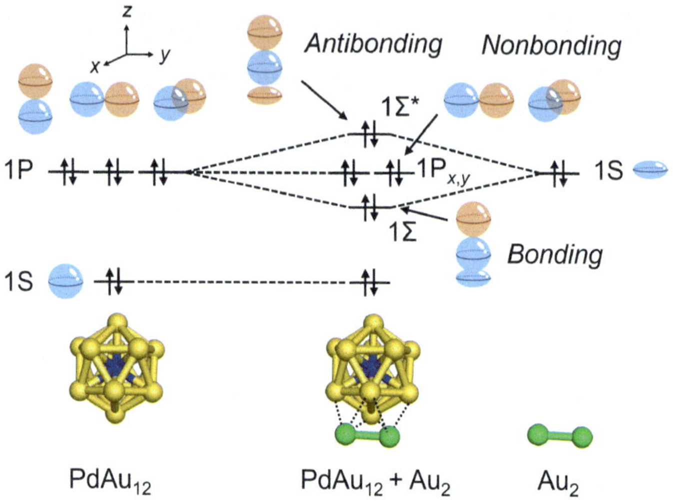

Scheme 6.1 Schematic illustration of the bonding diagram of $\mathbf{PdAu_{12}{-}2a^{2-}}$

1S orbital can be found at HOMO–3 (Fig. 6.7b). The bonding diagram between the $PdAu_{12}$ core with 8 electrons and the Au_2 unit with 2 electrons is illustrated in Scheme 6.1. $\mathbf{PdAu_{12}{-}2a^{2-}}$ can be categorized as a heteronuclear superatomic molecule composed of cores with different electrons and connected via the supervalence bonding between $(1S)^2$ and $(1S)^2(1P)^6$. The resulting electron configuration can be written as $(1S)^2(1\Sigma)^2(1P_{x,y})^4(1\Sigma^*)^2$ where 1S, $1P_x$, and $1P_y$ orbitals at the $PdAu_{12}$ core present as nonbonding superatomic orbitals. The nonbonding orbitals are originating from the heteronuclear nature as in the case of conventional diatomic molecules such as HCl. The nonbonding superatomic orbitals are corresponding to lone pairs in usual molecules, while its nature and property is the future subjects. This bonding diagram was also observed in the single point calculation of the bare $PdAu_{14}^{4+}$ core extracted from $\mathbf{PdAu_{12}{-}2a^{2-}}$ (Fig. 6.9), indicating that the structure of the core regulated the electronic structure. It is noteworthy that the bond order of $\mathbf{PdAu_{12}{-}2a^{2-}}$ was zero and correspond to the dimer of rare gas atoms like HeNe, but $\mathbf{PdAu_{12}{-}2a^{2-}}$ may be stabilized due to the interaction between the superatomic orbitals and d orbitals of Au atoms.

6.3.3 *Structure of $[PdAu_{24}(C{\equiv}CAr^F)_{18-2n}]^{2-}$ with More Than 10 Electrons*

It was found that $\mathbf{PdAu_{12}{-}2^{2-}}$ can be viewed as novel heteronuclear superatomic molecules composed of 8 electrons and 2 electrons. In this section, I investigated the applicability of this concept to other fragments by using DFT calculation, while

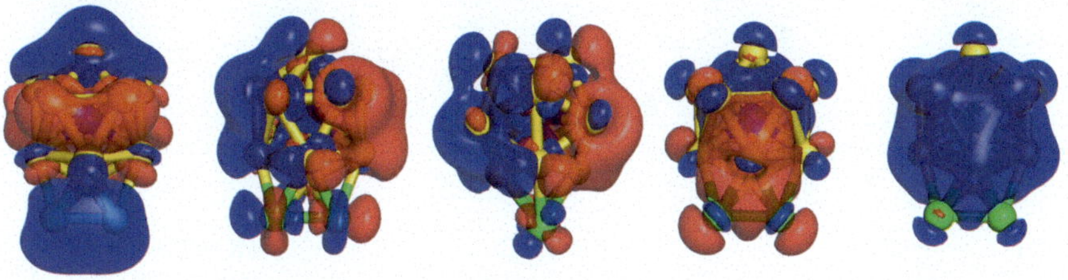

Fig. 6.9 Kohn–Sham orbitals obtained by single point calculation of $PdAu_{14}^{4+}$ extracted from $\mathbf{PdAu_{12}{-}2a^{2-}}$. Two percentages show the contribution of sp orbitals of the $PdAu_{12}$ core and that of the Au_2 unit, respectively

I did not aim at the structural determination of fragments because of the numerous possible isomers.

I conducted the DFT calculation on $\mathbf{PdAu_{12}{-}12^{2-}}$(20e) because it may take the highly stable closed-shell electron configuration with $(1S)^2(1P)^6(1D)^{10}(2S)^2$ as in the case of Au_{20} [47] or Al_7^+ [48] Such electronic structures can be identified from the delocalized valence electrons and the shape of molecular orbitals. An initial structure for $\mathbf{PdAu_{12}{-}12^{2-}}$(20e) was constructed by repeating the constructing procedure used for $\mathbf{PdAu_{12}{-}2a^{2-}}$ six times: two ligands coordinating to the same Au atom were removed from each $Au_2(C{\equiv}CAr^F)_3$ oligomer. The repeating removal of ligands resulted in the symmetric optimized structure referred to as $\mathbf{PdAu_{12}{-}12a^{2-}}$ in Fig. 6.10a. $\mathbf{PdAu_{12}{-}12a^{2-}}$ can be divided into the icosahedral $PdAu_{12}$ core and six $Au_2(C{\equiv}CAr^F)_1$ units parallel to each other. The CSM value of the $PdAu_{12}$ core was 0.088, indicating that the core is more symmetric than that of $\mathbf{PdAu_{12}{-}2a^{2-}}$ probably due to the less steric hindrance in the ligand layer of $\mathbf{PdAu_{12}{-}12a^{2-}}$. The VDE of $\mathbf{PdAu_{12}{-}12a^{2-}}$ was 2.10 eV and smaller than those of $\mathbf{PdAu_{12}{-}2a^{2-}}$ and $\mathbf{PdAu_{12}{-}0a^{2-}}$, which is consistent with the PE spectra (Fig. 6.3). Contribution analysis of atomic orbitals to molecular orbitals showed that only three orbitals from HOMO to HOMO−100 of $\mathbf{PdAu_{12}{-}12a^{2-}}$ had the contribution of sp orbitals of the $PdAu_{12}$ core greater than 10%. Such orbitals of $\mathbf{PdAu_{12}{-}12a^{2-}}$ are shown in Fig. 6.10b. HOMO and HOMO−1 can be assigned to 1P orbitals at the $PdAu_{12}$ core from its shape and HOMO−4 was composed of the Au_2 units and 1P orbitals, while I could not find 1D or 2S orbitals expected in the spherical jellium model. The Au_2 units made the complex orbitals like HOMO−7, HOMO−9, and HOMO−10 and it was difficult to assign them to the superatomic orbitals or supervalence bonding. Therefore, I concluded that $\mathbf{PdAu_{12}{-}12a^{2-}}$(20e) is not a single superatom with 20 electrons but the assembly of the $PdAu_{12}$ core with 8 electrons and six Au_2 units with 2 electrons. The electron configuration may be formally viewed as He_6Ne but it is unrealistic in the molecular orbital theory. More sophisticated theoretical model is required to interpret such a heteronuclear assembly of superatoms.

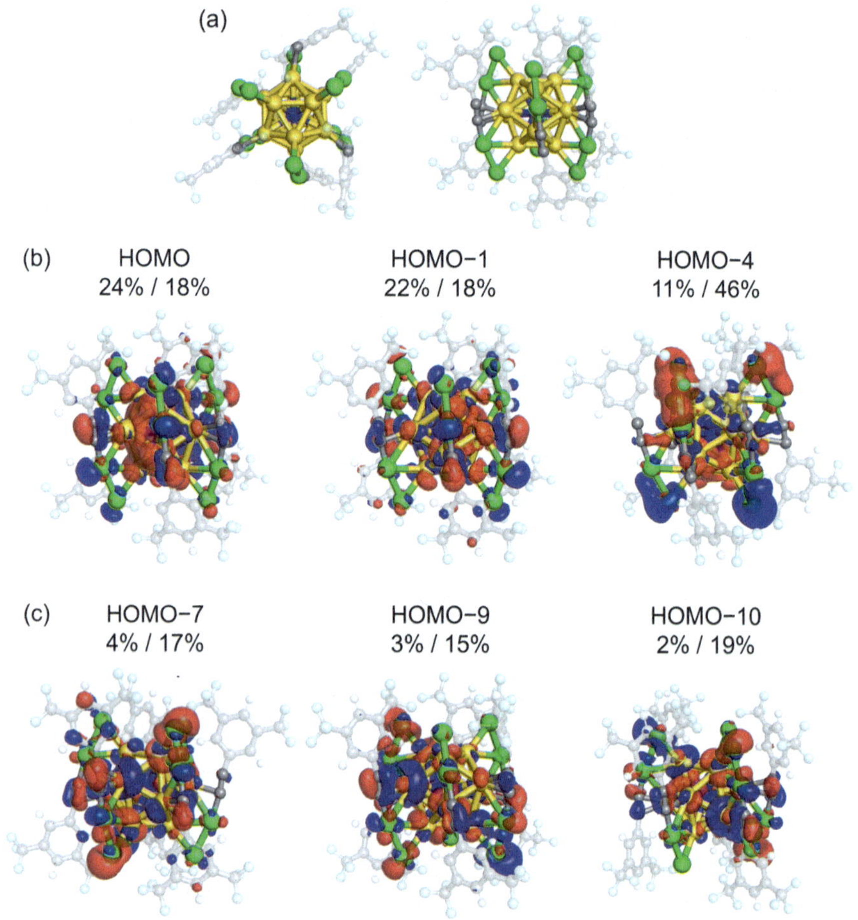

Fig. 6.10 **a** The optimized structure and **b** Kohn–Sham orbitals mainly composed from the $PdAu_{12}$ core and **c** the Au_2 units of $\mathbf{PdAu_{12}-12a^{2-}}$. Two percentages show the contribution of sp orbitals of the $PdAu_{12}$ core and that of the Au_2 unit, respectively

6.3.4 Mechanistic Insight into the CID Process

The heteronuclear superatomic molecules and heteronuclear assemblies of superatoms were observed as the fragment of $[PdAu_{24}(C\equiv CAr^F)_{18}]^{2-}$ by the successive collision-induced reductive elimination of 1,3-diynes. In this section I discussed the possible reason why the unique structure was generated based on the comparison of other dissociation processes.

It was reported that matrix-assisted laser desorption ionization (MALDI) MS of an alkynyl-protected cluster $Au_{22}(C\equiv CR)_{18}$ resulted in the efficient desorption of

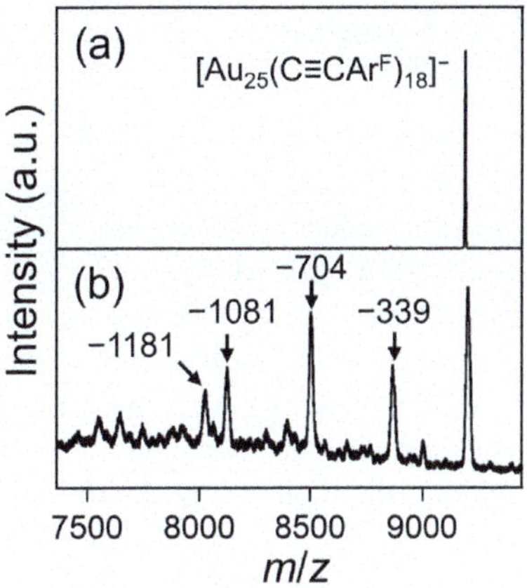

Fig. 6.11 Negative mode MALDI mass spectra of $[Au_{25}(C{\equiv}CAr^F)_{18}]^-$ at **a** a threshold laser fluence **b** a high laser fluence. Numbers in panel **b** show the decrease of mass compared from $[Au_{25}(C{\equiv}CAr^F)_{18}]^-$

$Au_4(C{\equiv}CR)_4$ to yield pure $Au_{18}(C{\equiv}CR)_{14}$ in the mass spectra while the desorption of 1,3-diyne was not observed [29]. MALDI MS of $[Au_{25}(C{\equiv}CAr^F)_{18}]^-$ was conducted instead of $[PdAu_{24}(C{\equiv}CAr^F)_{18}]^{2-}$ because MALDI MS did not work well with multiply charged ions in general. The mass spectrum at the high laser fluence (Fig. 6.11b) did not exhibit the loss of $(C{\equiv}CAr^F)_2$ (M.W. = 474) or $Au_4(C{\equiv}CAr^F)_4$ (M.W. = 1736) but the complicated dissociation other than $Au_x(C{\equiv}CAr^F)_y$. Although the fragmentation in both CID MS and MALDI MS is the thermal process, the difference in the fragmented product could be raised from the time scale. In CID MS, ions are gradually heated by the numerous numbers of collisions with the gas molecules for several μs and will result in the kinetically controlled dissociation by passing the lowest energy barrier. In MALDI MS, ions are gently but rapidly evaporated by the nanosecond laser and thus the ions can be excited to higher states and the fragmentation will proceed thermodynamically.

In structurally similar thiolate-protected cluster $[Au_{25}(SR)_{18}]^-$, both CID MS and MALDI MS resulted in the desorption of a neutral complex $Au_4(SR)_4$ [8, 49]. The different dissociation process between alkynyls and thiolates is possibly due to the difference in the energy gain by the C–C or S–S bond formation. DFT calculation was conducted to calculate the bonding energy between two $C{\equiv}CAr^F$ and two $2\text{-}PhC_2H_4S$. The C–C bond formation between two $C{\equiv}CAr^F$ ligands was exothermic by 7.27 eV while S–S bond formation between two $2\text{-}PhC_2H_4S$ was only 2.70 eV. As a result, the reductive elimination process $\mathbf{PdAu_{12}\text{–}0a^{2-}} \rightarrow \mathbf{PdAu_{12}\text{–}2a^{2-}} + Ar^FC{\equiv}C\text{–}C{\equiv}CAr^F$ was endothermic by 0.83 eV, while the similar process for thiolate-protected clusters might be endothermic by several eV.

These considerations suggested that the kinetically controlled desorption of ligands driven by the C–C bond formation seems to be a key to generate heteronuclear assemblies of superatoms. Although the kinetically controlled desorption happened in the gas-phase, the mechanistic insights into the formation process of the heteronuclear assemblies of superatoms will contribute to the strategy of wet chemical synthesis of such clusters.

6.4 Summary

In summary, I clarified the binding mechanism of excess electrons in $[PdAu_{24}(C{\equiv}CAr^F)_{18-2n}]^{2-}$ with $8 + 2n$ electrons generated by the collision-induced reductive elimination of 1,3-diynes from an eight-electron superatom $[PdAu_{24}(C{\equiv}CAr^F)_{18}]^{2-}$. A joint study of photoelectron spectroscopy and density functional theory calculation showed that the simplest product $[PdAu_{24}(C{\equiv}CAr^F)_{16}]^{2-}$ with 10 electrons was composed of the $PdAu_{12}$ core with 8 electrons and the Au_2 unit with 2 electrons rather than a single superatom with 10 electrons. The segregated electronic structure can be viewed as a heteronuclear superatomic molecule and will be valid even at $n = 6$. This study not only reveals the unique binding mechanism of valence electrons in the ligand-protected clusters generated by CID but also will be the basis of the bonding theory of heteronuclear superatomic molecules.

References

1. Papayannopoulos IA (1995) Mass Spectrom Rev 14:49
2. Johnson AR, Carlson EE (2015) Anal Chem 87:10668
3. Cooks RG, Patrick JS, Kotiaho T, McLuckey SA (1994) Mass Spectrom Rev 13:287
4. Rijs N, Khairallah GN, Waters T, O'Hair RAJ (2008) J Am Chem Soc 130:1069
5. Waters T, O'Hair RAJ, Wedd AG (2003) Int J Mass Spectrom 228:599
6. Tsukuda T, Häkkinen H (2015) Protected metal clusters: from fundamentals to applications, 1st edn. Elsevier B. V, Amsterdam
7. Jin R, Zeng C, Zhou M, Chen Y (2016) Chem Rev 116:10346
8. Angel LA, Majors LT, Dharmaratne AC, Dass A (2010) ACS Nano 4:4691
9. Black DM, Bhattarai N, Whetten RL, Bach SBH (2014) J Phys Chem A 118:10679
10. Ghosh A, Bodiuzzaman M, Nag A, Jash M, Baksi A, Pradeep T (2017) ACS Nano 11:11145
11. Chakraborty P, Baksi A, Khatun E, Nag A, Ghosh A, Pradeep T (2017) J Phys Chem C 121:10971
12. Jash M, Reber AC, Ghosh A, Sarkar D, Bodiuzzaman M, Basuri P, Baksi A, Khanna SN, Pradeep T (2018) Nanoscale 10:15714
13. Tomihara R, Hirata K, Yamamoto H, Takano S, Koyasu K, Tsukuda T (2018) ACS Omega 3:6237
14. Baksi A, Jash M, Bag S, Mudedla SK, Bodiuzzaman M, Ghosh D, Paramasivam G, Subramanian V, Pradeep T (2019) J Phys Chem C 123:28494
15. Narouz MR, Osten KM, Unsworth PJ, Man RWY, Salorinne K, Takano S, Tomihara R, Kaappa S, Malola S, Dinh C-T, Padmos JD, Ayoo K, Garrett PJ, Nambo M, Horton JH, Sargent EH, Häkkinen H, Tsukuda T, Crudden CM (2019) Nat Chem 11:419
16. Hirata K, Tomihara R, Kim K, Koyasu K, Tsukuda T (2019) Phys Chem Chem Phys 21:17463
17. Heaven MW, Dass A, White PS, Holt KM, Murray RW (2008) J Am Chem Soc 130:3754
18. Zhu M, Aikens CM, Hollander FJ, Schatz GC, Jin R (2008) J Am Chem Soc 130:5883
19. Tang Q, Jiang D-E (2015) J Phys Chem C 119:10804
20. Tang Q, Jiang D-E (2017) Chem Mater 29:6908
21. Shichibu Y, Konishi K (2010) Small 6:1216
22. Li J-J, Guan Z-J, Lei Z, Hu F, Wang Q-M (2019) Angew Chem Int Ed 58:1083
23. Narouz MR, Takano S, Lummis PA, Levchenko TI, Nazemi A, Kaappa S, Malola S, Yousefalizadeh G, Calhoun LA, Stamplecoskie KG, Häkkinen H, Tsukuda T, Crudden CM (2019) J Am Chem Soc 141:14997

24. Shen H, Xiang S, Xu Z, Liu C, Li X, Sun C, Lin S, Teo BK, Zheng N (2020) Nano Res 12:13
25. Walter M, Akola J, Lopez-Acevedo O, Jadzinsky PD, Calero G, Ackerson CJ, Whetten RL, Grönbeck H, Häkkinen H (2008) Proc Natl Acad Sci 105:9157
26. de Heer WA (1993) Rev Mod Phys 65:611
27. Jena P, Sun Q (2018) Chem Rev 118:5755
28. Lei Z, Wan X-K, Yuan S-F, Guan Z-J, Wang Q-M (2018) Acc Chem Res 51:2465
29. Ito S, Takano S, Tsukuda T (2019) J Phys Chem Lett 10:6892
30. Han X-S, Luan X, Su H-F, Li J-J, Yuan S-F, Lei Z, Pei Y, Wang Q-M (2020) Angew Chem Int Ed 59:2309
31. Guan Z-J, Hu F, Li J-J, Wen Z-R, Lin Y-M, Wang Q-M (2020) J Am Chem Soc 142:2995
32. Takano S, Ito S, Tsukuda T (2019) J Am Chem Soc 141:15994
33. Ito S, Koyasu K, Takano S, Tsukuda T (2020) J Phys Chem C 124:19119
34. Brandys M-C, Jennings MC, Puddephatt RJ (2000) J Chem Soc Dalton Trans 4601
35. Takano S, Hirai H, Muramatsu S, Tsukuda T (2018) J Am Chem Soc 140:8380
36. Takano S, Hirai H, Muramatsu S, Tsukuda T (2018) J Am Chem Soc 140:12314
37. Frisch MJ, Trucks GW, Schlegel HB, Scuseria GE, Robb MA, Cheeseman JR, Scalmani G, Barone V, Petersson GA, Nakatsuji H, Li X, Caricato M, Marenich AV, Bloino J, Janesko BG, Gomperts R, Mennucci B, Hratchian HP, Ortiz JV, Izmaylov AF, Sonnenberg JL, Young WD, Ding F, Lipparini F, Egidi F, Goings J, Peng B, Petrone A, Henderson T, Ranasinghe D, Zakrzewski VG, Gao J, Rega N, Zheng G, Liang W, Hada M, Ehara M, Toyota K, Fukuda R, Hasegawa J, Ishida M, Nakajima T, Honda Y, Kitao O, Nakai H, Vreven T, Throssell K, Montgomery Jr. JA, Peralta JE, Ogliaro F, Bearpark MJ, Heyd JJ, Brothers EN, Kudin KN, Staroverov VN, Keith TA, Kobayashi R, Normand J, Raghavachari K, Rendell AP, Burant JC, Iyengar SS, Tomasi J, Cossi M, Millam JM, Klene M, Adamo C, Cammi R, Ochterski JW, Martin RL, Morokuma K, Farkas O, Foresman JB, Fox DJ (2016) Gaussian 16, Revision B.01. Gaussian, Inc., Wallingford CT
38. O'Boyle NM, Tenderholt AL, Langner KM (2008) J Comput Chem 29:839
39. Glendening ED, Reed AE, Carpenter JE, Weinhold F, NBO, Ver. 3.1
40. Ito S, Koyasu K, Takano S, Tsukuda T (2021) J Phys Chem Lett 12:10417
41. Ito S, Tasaka Y, Nakamura K, Fujiwara Y, Hirata K, Koyasu K, Tsukuda T (2022) J Phys Chem Lett 13:5049
42. Wang L-S, Wang X-B (2000) J Phys Chem A 104:1978
43. Ok KM, Halasyamani PS, Casanova D, Llunell M, Alemany P, Alvarez S (2006) Chem Mater 18:3176
44. Tofanelli MA, Salorinne K, Ni TW, Malola S, Newell B, Phillips B, Häkkinen H, Ackerson CJ (2016) Chem Sci 7:1882
45. Qian H, Eckenhoff WT, Zhu Y, Pintauer T, Jin R (2010) J Am Chem Soc 132:8280
46. Cheng L, Ren C, Zhang X, Yang J (2013) Nanoscale 5:1475
47. Li J, Li X, Zhai H-J, Wang L-S (2003) Science 299:864
48. Jarrold MF, Bower JE (1986) J Chem Phys 85:5373
49. Dass A, Stevenson A, Dubay GR, Tracy JB, Murray RW (2008) J Am Chem Soc 130:5940

Chapter 7
Summary and Prospects

Abstract This chapter summarizes my thesis and provides prospects for superatom chemistry. The thesis described the effect of structural factors and the bonding interaction on structurally defined superatoms. The energy level of 1P orbitals can be tuned coarsely by the heterometal doping and finely by the stepwise ligand exchange, respectively. The electronic structure of both homonuclear and heteronuclear superatomic molecules was examined by anion PES and to develop the bonding theory of superatoms. This thesis will provide fundamental insights and design principles of structurally defined superatoms and stimulate the future research on their chemical and physical properties.

Keywords Superatom · Superatomic physics · Photoelectron spectroscopy · Metal clusters

7.1 Summary

In this thesis, I studied the superatomic electronic structure and the bonding interaction of structurally defined superatoms by using gas-phase anion photoelectron spectroscopy (PES) and density functional theory (DFT) calculation to understand the intrinsic electronic structure and to develop the design principle of superatoms. These results were rationalized by using a superatom concept, a jellium potential, and a super valence bond (SVB) model.

In Chap. 2, I described the experimental setup of anion PES and especially the detail of the quadrupole linear ion trap. The ion trap successfully enhanced the ion intensity and the photoelectron counts about two orders of magnitude, resulting in the significant improvement of the efficiency and the quality of anion PES.

In Chap. 3, I investigated the effect of the heterometal doping by anion PES on thiolate-protected superatoms $[Au_{25}(PET)_{18}]^-$(8e) (PET = 2-PhC_2H_4S) and $[MAg_{24}(DMBT)_{18}]^z$(8e) (DMBT = 2,4-$(CH_3)_2C_6H_3$; $z = -1$ for M = Ag and Au; $z = -2$ for M = Pd and Pt). Laser-fluence-dependent PE spectra and pump–probe PES on a significantly intensified mass-selected ion beam showed that the

S. Ito, *Electronic Structures and Bonding Interaction of Structurally Defined Gold/Silver Superatoms*, Springer Theses, https://doi.org/10.1007/978-981-96-2892-6_7

previous PE spectra were contaminated by PE signals due to the two-photon electron detachment process via long-lived photoexcited states. The energy level of 1P orbitals of the Ag_{13} superatom with respect to the vacuum level (~4 eV) was not appreciably affected by doping an Au atom at the center, but was reduced by ~2 eV by doping Pd or Pt. The large doping effect of group 10 elements was rationalized as the upward shift of the jellium potential by an increased negative charge and the two-step jellium potential due to the different valency between Ag and Pd.

In Chap. 4, I investigated the effect of the stepwise ligand substitution by anion PES on $[PdAu_{24}(C{\equiv}CAr^F)_{18-x}(C{\equiv}CPh)_x]^{2-}$(8e) ($Ar^F$ = 3,5-$(CF_3)_2C_6H_3$) where the electron-withdrawing Ar^F moiety were thought to play an essential role in the stabilization. Anion PES coupled with mass spectrometer revealed the electron binding energy with specific x ranging from 0 to 6. I found that electron binding energies decreased linearly with x, indicating that the electron-withdrawing CF_3 substituents on the alkynyl ligand played a critical role in the electronic stabilization of $[PdAu_{24}(C{\equiv}CAr^F)_{18}]^{2-}$. DFT calculations reproduced the decrease of electron binding energies and rationally explained the ligand effect as the modulation of the depth of the jellium potential by the electrostatic effect of the dipole moment of the ligand.

In Chap. 5, I validated an SVB model in a homonuclear superatomic molecule $Au_{38}(PET)_{24}$(14e). The SVB model rationalized the electronic structure of a bi-icosahedral Au_{23} core as $(1\Sigma)^2(1\Sigma^*)^2(1\Pi)^4(2\Sigma)^2(1\Pi^*)^4$ where each orbital is created by the bonding and antibonding interactions between two 1S and 1P superatomic orbitals. To validate the model, I conducted anion PES on $[MAu_{37}(PET)_{24}]^-$ (14e) (M = Pd, Pt) which are isoelectronic and isostructural with $Au_{38}(PET)_{24}$. The PE spectra of $[M_1Au_{37}(PET)_{24}]^-$ exhibited two distinctive peaks as predicted by the SVB model: one from the nearly-degenerate $1\Pi^*$ orbitals and the other from nearly-degenarate 1Π and 2Σ orbitals.

In Chap. 6, I investigated the binding mechanism of excess electrons in $[PdAu_{24}(C{\equiv}CAr^F)_{18-2n}]^{2-}$($(8 + 2n)$e) ($n$ = 1–6) with $8 + 2n$ electrons generated by the collision-induced reductive elimination of 1,3-diynes from $[PdAu_{24}(C{\equiv}CAr^F)_{18}]^{2-}$(8e) in the gas phase. In the simplest fragment with 10 electrons, the increased two electrons were accommodated by the Au_2 unit initially coordinated by the desorbed ligands while the $PdAu_{12}$ core retained the closed-shell 8-electron configuration. Based on the SVB model, the 10-electron product can be viewed as an unprecedented heteronuclear superatomic molecule with the electron configuration of $(1S)^2(1\Sigma)^2(1P_{x,y})^4(1\Sigma^*)^2$.

In conclusion, the effect of structural factors and the bonding interaction on structurally defined superatoms is summarized in Fig. 7.1. The energy level of 1P orbitals can be tuned coarsely by the heterometal doping and finely by the stepwise ligand exchange, respectively. The electronic structure of both homonuclear and heteronuclear superatomic molecules was examined by anion PES and to develop the bonding theory of superatoms.

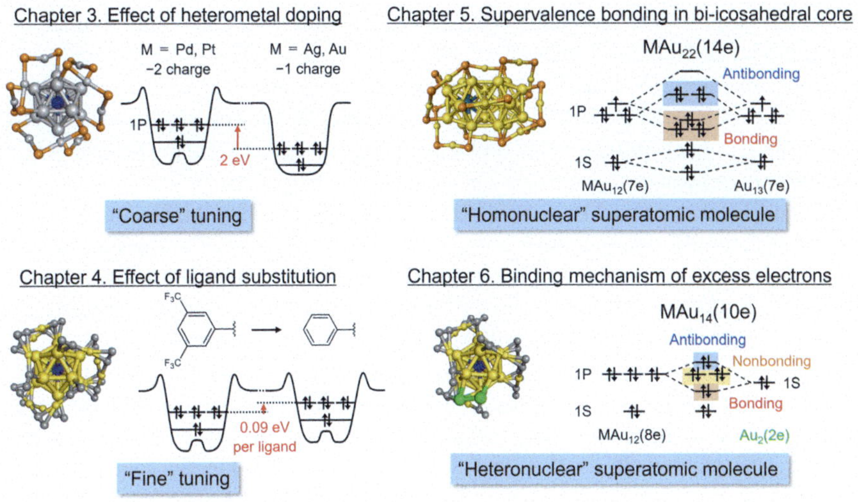

Fig. 7.1 Graphical summary of my doctoral thesis

7.2 Future Prospects

My doctoral study based on anion PES provided fundamental insights into the electronic structures of structurally defined superatoms. In this section, I described the potential application of my works for the structurally defined superatoms and the prospects of anion PES of structurally defined superatoms.

It was found that the energy level of superatomic 1P orbitals located at highest occupied molecular orbitals (HOMOs) can be tuned by ~2 eV by doping group 10 elements in Chap. 3 and by ~70 meV by a single ligand exchange in Chap. 4. These two tuning principles correspond to the "coarse" and "fine" tuning of the energy level and the combination of two strategies will enable the precise control of the 1P orbitals for the specific application such as photocatalyst [1], electrocatalyst, and electrochemical sensing [2], where the energy-level matching between the catalyst or the sensor and the reactant is critical to enhance the activity.

Structurally defined superatoms with unique physicochemical properties are promising building units of functional nanomaterials. The SVB model validated in Chap. 5 and extended to the heteronuclear structure in Chap. 6 will provide a general and comprehensive understanding of the electronic structure of structurally defined superatomic molecules and further assemblies. It is noteworthy that the heteronuclear superatomic molecules discussed in Chap. 6 have not been synthesized by a wet chemical synthesis. It will be necessary to develop a new strategy to obtain such heteronuclear structures instead of the direct reduction method relying on the serendipity. One promising approach is the controlled dimerization of superatoms where a 6-electron superatom $[MAu_8(PPh_3)_8]^{2+}$ (M = Pd, Pt) and a 7-electron superatom $[M'Au_{24}(SR)_{18}]^-$ were reacted to obtain a 12-electron superatomic molecule $MM'Au_{36}(SR)_{24}$ [3]. The usage of 2-electron superatom such as $[Au_4(PPh_3)_4]^{2+}(2e)^4$ instead of the 6-electron superatom may yield the heteronuclear superatomic molecule.

Anion PES was proved to be a powerful method to evaluate the electronic structure of structurally defined superatoms in my doctoral thesis. On the other hand, mass spectrometry has been an indispensable tool for the chemistry of structurally defined superatoms to evaluate the composition and to understand the reaction mechanism [5]. Therefore, Anion PES will be applicable not only to evaluate the intrinsic electronic structure of structurally defined superatoms, but also to understand the mechanism of synthesis or complexation. For example, the time-course mass spectrometry of the synthesis of $[Au_{25}(SR)_{18}]^-$ revealed the presence of various intermediate clusters with 0, 2, 4, 6, and 8 electrons [6]. The kinetic study of intercluster metal-exchange reaction between $[Au_{25}(SR)_{18}]^-$ and $[Ag_{25}(SR)_{18}]^-$ by using mass spectrometry detected and confirmed the presence of the dimeric complex $[Au_{25}(SR)_{18}{\cdot}Au_{25}(SR)_{18}]^{2-}$ as intermediates [7]. These mass spectrometric studies demonstrated the generation of unique intermediates in solution, while further characterization of these species is still challenging. Anion PES can shed light on the electronic structure of these intermediates and give a comprehensive view of the mechanism of the clustering process and the intercluster reaction.

PES is applicable to any molecule regardless of its charge states. PES of cationic or neutral ligand-protected clusters will significantly increase the scope of clusters because most of them are not anions. One of the fundamental questions about metal clusters is identifying the critical size at which its electronic structure changes from non-metallic to metallic [8]. It has been investigated in solution phase by using linear and transient absorption spectroscopy and electrochemical measurement. According to the previous study of bare metal clusters [9], nonmetal-to-metal transition can be probed by PES using VUV light because the size-dependent screening of core-ionized neutral cluster can serve as a sensor. PES using VUV light may also reveal the location of the deep 1S orbital.

Structurally defined gold/silver superatoms possess discrete energy levels similar to atomic orbitals. Therefore, they are expected to mimic the properties of conventional elements in the periodic table and to work as artificial elements. One of the mimicking aspects is the high stability at the closed-shell electron configuration [10]. The super valence bond model of superatoms discussed in Chaps. 5 and 6 is an analogy to the molecular orbital theory. I think that there should be plenty of room for the exploration of the physics of structurally defined superatoms. One physical phenomenon is the spin–orbit splitting: relativistic density functional theory calculation showed that the superatomic 1P orbitals are split into two by the spin–orbit coupling, which is analogous to the spin–orbit coupling in halogen atoms to yield $^2P_{1/2}$ and $^2P_{3/2}$ [11]. Anion PES will be applicable to determine the split energy experimentally and elucidate the effect of spin–orbit coupling on the property. The field of atomic physics provides several questions for superatoms. For example, is it possible to make and characterize the Rydberg state of superatoms with a high principle quantum number? Are there any effects on the energy level of superatomic orbitals under the electric field (Stark effect) or the magnetic field (Zeeman effect)? How about the hyperfine structure of superatomic orbitals? These questions will pave the way for "superatomic physics" to understand the fundamental laws of structurally defined superatoms in comparison with atoms and will maximize the potential of the superatoms as controllable quantum units of functional materials.

In conclusion, this thesis will provide fundamental insights and design principles of structurally defined superatoms and stimulate the future research on their chemical and physical properties.

References

1. Hirai H, Takano S, Nakashima T, Iwasa T, Taketsugu T, Tsukuda T (2022) Angew Chem Int Ed 61:e202207290
2. Kwak K, Lee D (2019) Acc Chem Res 52:12–22
3. Ito E, Takano S, Nakamura T, Tsukuda T (2021) Angew Chem Int Ed 60:645
4. Zhang C, Ding M, Ren Y, Ma A, Yin Z, Ma X, Wang S (2023) Nanoscale Adv 5:3287
5. Chen T, Yao Q, Nasaruddin RR, Xie J (2019) Angew Chem Int Ed 58:11967
6. Chen T, Fung V, Yao Q, Luo Z, Jiang D, Xie J (2018) J Am Chem Soc 140:11370
7. Neumaier M, Baksi A, Weis P, Schneider EK, Chakraborty P, Hahn H, Pradeep T, Kappes MM (2021) J Am Chem Soc 143:6969
8. Zhou M, Du X, Wang H, Jin R (2021) ACS Nano 15:13980
9. Senz V, Fischer T, Oelßner P, Tiggesbäumker J, Stanzel J, Bostedt C, Thomas H, Schöffler M, Foucar L, Martins M et al (2009) Phys Rev Lett 102:138303
10. Walter M, Akola J, Lopez-Acevedo O, Jadzinsky PD, Calero G, Ackerson CJ, Whetten RL, Gronbeck H, Hakkinen H (2008) Proc Natl Acad Sci 105:9157
11. Jiang D, Kühn M, Tang Q, Weigend F (2014) J Phys Chem Lett 5:3286

Curriculum Vitae

Shun Ito

Department of Chemistry, Graduate School of Science, The University of Tokyo (UTokyo)

7-3-1 Hongo, Bunkyo-ku, Tokyo, 113-0033, Japan

E-mail: s.ito@chem.s.u-tokyo.ac.jp

Web: https://sites.google.com/view/shun-ito/home

Appointments

Project Researcher The University of Tokyo	Apr. 2024–Current
JSPS research fellowship for young scientists (DC1) Japan Society for the Promotion of Science	Apr. 2021–Mar. 2024

S. Ito, *Electronic Structures and Bonding Interaction of Structurally Defined Gold/Silver Superatoms*, Springer Theses, https://doi.org/10.1007/978-981-96-2892-6

TAKUETSU (Excellence) Research Assistant Forefront Physics and Mathematics Program to Drive Transformation (FoPM) The University of Tokyo	Sept. 2019–Mar. 2021

Education

Doctor of Philosophy (Ph.D.) in Science The University of Tokyo, Japan Supervised by Prof. Dr. Tatsuya Tsukuda	Mar. 2024
Master of Science (MSc) The University of Tokyo, Japan Supervised by Prof. Dr. Tatsuya Tsukuda	Mar. 2021
Bachelor of Science (B. Sc.) The University of Tokyo, Japan Supervised by Prof. Dr. Tatsuya Tsukuda	Mar. 2019

Honors and Awards

The Graduate School of Science Encouragement Award, UTokyo	Mar. 2024
Excellent Award, FoPM	Sep. 2023
CSJ Student Presentation Award, Chemical Society of Japan	Mar. 2023
Young best presentation award, Society of Nano Science and Technology	May 2021
The Graduate School of Science Encouragement Award, UTokyo	Mar. 2021
The School of Science Encouragement Award, UTokyo	Mar. 2019
Department of Chemistry Award, UTokyo	Mar. 2019

Research Interests

My research aims to understand the unique physicochemical properties of ligand-protected gold superatoms, where clusters of gold atoms behave like atoms. Their properties intimately correlate with the discrete superatomic electronic structures formed by valence electrons of gold atoms. I investigated the unique electronic structure of ligand-protected superatoms by gas-phase anion photoelectron spectroscopy (PES). I developed the highly efficient setup of PES and revealed how the valence electrons of gold clusters are affected by the structural parameters. I am also interested in the gas-phase collisional excitation as a unique method to investigate intrinsic stability of ligand-protected gold clusters and to explore novel superatomic compounds.

Publications

1. Resistance of a $PdAu_{12}$(8e) Core to Growth in Collision-Induced Sequential Reductive Elimination of $(C\equiv CR)_2$ from $[PdAu_{24}(C\equiv CR)_{18}]^{2-}$
   Shun Ito, Koto Hirano, Kiichirou Koyasu, Xian-Kai Wan, Quan-Ming Wang, and Tatsuya Tsukuda*
   J. Phys. Chem. Lett. **15**, 11060 (2024).

2. Revisiting the structure of $[PdAu_9(PPh_3)_8(CN)]^{2+}$ produced by atmospheric pressure plasma irradiation of $[PdAu_8(PPh_3)_8]^{2+}$ in methanol
Takumi Imagawa, **Shun Ito**, Frank Hennrich, Marco Neumaier, Patrik Weis, Kiichirou Koyasu, Manfred M. Kappes*, Tatsuya Tsukuda*
J. Chem. Phys. **161**, 024303 (2024).
3. Gas-Phase Anion Photoelectron Spectroscopy of Alkanethiolate-Protected $PtAu_{12}$ Superatoms: Charging Energy in Vacuum vs Solution
Yuriko Tasaka, Megumi Suyama, **Shun Ito**, Kiichirou Koyasu, Manfred Kappes*, Flavio Maran*, Tatsuya Tsukuda*
Angew. Chem. Int. Ed. **63**, e202408335 (2024).
4. Gas-phase Structures of $[Au_{21}(SR)_{14}]^-$ and $[Au_{17}(SR)_{10}]^-$ with Eight Electrons: Can They Support an Icosahedral Au_{13} Core?
Yuki Fujiwara, **Shun Ito**, Kiichirou Koyasu, Tatsuya Tsukuda*
J. Phys. Chem. A **128**, 3119 (2024).
5. Cyclic ion mobility of doped $[MAu_{24}L_{18}]^{2-}$ superatoms and their fragments (M = Ni, Pd and Pt; L = alkynyl)
Frank Hennrich, **Shun Ito**, Patrick Weis, Marco Neumaier, Shinjiro Takano, Tatsuya Tsukuda, and Manfred M. Kappes*
Phys. Chem. Chem. Phys. **26**, 8408 (2024).
6. Vibrationally Resolved Absorption, Fluorescence, and Preresonance Raman Spectroscopy of Isolated Pyronin Y Cation at 5 K
Sreekanta Debnath*, Alexander Schäfer, **Shun Ito**, Dmitry Strelnikov, Rabea Schneider, Karolina A. Haupa, and Manfred M. Kappes*
J. Phys. Chem. Lett. **14**, 10553 (2023).
7. Effect of total charge on the electronic structure of thiolate-protected $X@Ag_{12}$ superatoms (X = Ag, Au)
Katsunosuke Nakamura, **Shun Ito**, Kiichirou Koyasu, and Tatsuya Tsukuda*
Phys. Chem. Chem. Phys. **25**, 5955 (2023).
8. Supervalence Bonding in Bi-icosahedral Cores of $[M_1Au_{37}(SC_2H_4Ph)_{24}]^-$ (M = Pd and Pt): Fusion-Mediated Synthesis and Anion Photoelectron Spectroscopy
Emi Ito†, **Shun Ito**†, Shinjiro Takano, Toshikazu Nakamura, and Tatsuya Tsukuda* (†: Equal contribution)
JACS Au **2**, 2627 (2022).
9. Electron Affinities of Ligated Icosahedral M_{13} Superatoms Revisited by Gas-Phase Anion Photoelectron Spectroscopy
Shun Ito, Yuriko Tasaka, Katsunosuke Nakamura, Yuki Fujiwara, Keisuke Hirata, Kiichirou Koyasu, and Tatsuya Tsukuda*
J. Phys. Chem. Lett. **13**, 5049 (2022).
10. Correction to "Photoinduced Thermionic Emission from $[M_{25}(SR)_{18}]^-$ (M = Au, Ag) Revealed by Anion Photoelectron Spectroscopy"
Keisuke Hirata, **Shun Ito**, Kuenhee Kim, Katsunosuke Nakamura, Hirokazu Kitazawa, Shun Hayashi, Kiichirou Koyasu, and Tatsuya Tsukuda*
J. Phys. Chem. C **126**, 8965 (2022).

11. Critical Role of CF_3 Groups in the Electronic Stabilization of $[PdAu_{24}(C{\equiv}C_6H_3(CF_3)_2)_{18}]^{2-}$ as Revealed by Gas-Phase Anion Photoelectron Spectroscopy
Shun Ito, Kiichirou Koyasu, Shinjiro Takano, and Tatsuya Tsukuda*
J. Phys. Chem. Lett. **12**, 10417 (2021).
12. Ligand-Protected Gold/Silver Superatoms: Current Status and Emerging Trends
Haru Hirai, **Shun Ito**, Shinjiro Takano, Kiichirou Koyasu and Tatsuya Tsukuda*
Chem. Sci. **11**, 12233 (2020).
13. Collision–Induced Reductive Elimination of 1,3–Diynes from $[MAu_{24}(C{\equiv}CR)_{18}]^{2-}$ (M = Pd, Pt) Yielding Clusters of Superatoms
Shun Ito, Kiichirou Koyasu, Shinjiro Takano, and Tatsuya Tsukuda*
J. Phys. Chem. C **124**, 19119 (2020).
14. Alkynyl-Protected $Au_{22}(C{\equiv}CR)_{18}$ Clusters Featuring New Interfacial Motifs and R-Dependent Photoluminescence
Shun Ito, Shinjiro Takano, and Tatsuya Tsukuda*
J. Phys. Chem. Lett. **10**, 6892 (2019).
15. Efficient and Selective Conversion of Phosphine-Protected $(MAu_8)^{2+}$ (M = Pd, Pt) Superatoms to Thiolate-Protected $(MAu_{12})^{6+}$ or Alkynyl-Protected $(MAu_{12})^{4+}$ Superatoms via Hydride Doping
Shinjiro Takano*, **Shun Ito**, and Tatsuya Tsukuda*
J. Am. Chem. Soc. **141**, 15994 (2019).

The manufacturer's authorised representative in the EU is Springer Nature Customer Service Centre GmbH, Europaplatz 3, 69115 Heidelberg, Germany. If you have any concerns regarding our products, please contact ProductSafety@springernature.com

Printed and bound by CPI Group (UK) Ltd, Croydon, CR0 4YY
01/10/2026
02225945-0006